Networks of Innovation

Networks of Innovation

Change and Meaning in the Age of the Internet

ILKKA TUOMI

OXFORD

UNIVERSITY PRESS

OXFORD

UNIVERSITY PRESS

Great Clarendon Street, Oxford OX2 6DP

Oxford University Press is a department of the University of Oxford.
It furthers the University's objective of excellence in research, scholarship,
and education by publishing worldwide in

Oxford New York

Auckland Bangkok Buenos Aires Cape Town Chennai
Dar es Salaam Delhi Hong Kong Istanbul Karachi Kolkata
Kuala Lumpur Madrid Melbourne Mexico City Mumbai Nairobi
São Paulo Shanghai Taipei Tokyo Toronto

Oxford is a registered trade mark of Oxford University Press
in the UK and in certain other countries

Published in the United States
by Oxford University Press Inc., New York

© SITRA Helsinki, 2002

Sitra's Publication Series, publication no. 249,
ISSN 0785-8388 (Sitra)
Sitra – the Finnish National Fund for Research and Development

British Library Cataloguing in Publication Data

Data available

Library of Congress Cataloging in Publication Data

Tuomi, Ilkka.
    Networks of innovation : change and meaning in the age of the Internet / Ilkka Tuomi.
        p. cm.
    Includes bibliographical references and indexes.
    1. Technological innovations. 2. Computer networks. 3. Internet. 4. Linux. I. Title.
    T173.8 T86 2002   303.48'3—dc21   2002070154

ISBN 0-19-925698-5

1 3 5 7 9 10 8 6 4 2

Typeset by Newgen Imaging Systems (P) Ltd., Chennai, India
Printed in Great Britain
on acid-free paper by
T. J. International Ltd., Padstow, Cornwall

ACKNOWLEDGEMENTS

Many friends and colleagues provided ideas, comments, and support while I was working on this book. I wrote the first version of the manuscript during my year-and-a-half stay at the University of California, Berkeley. Sitra, the Finnish National Fund for Research and Development, kindly supported my research at Berkeley, as a part of their innovation research program. My former employer Nokia provided partial funding for my research project.

Finished books are often the result of years of work and they can contain many layers of ideas built one over another. It may be difficult to acknowledge all those intellectual debts that go into writing a book. Sometimes, however, it is easy to recognize the importance of a specific individual. Without the personal and intellectual support of Manuel Castells I could not have written this book. Manuel invited me to Berkeley, commented on my work and the manuscript at different phases, and, together with Emma Kiselyova, helped me to feel at home there.

Berkeley is one of the intellectual hot spots in the world. During my stay there, I was able to discuss my ideas in various seminars, workshops, and conferences, and greatly enjoyed the opportunity to talk with many brilliant thinkers and researchers. I would like to thank in particular John Canny and Jerry Feldman, who both gave feedback on my work and linked me with many key researchers in the San Francisco Bay area. Jerry also organized excellent working conditions for me at the International Computer Science Institute. I gained much from discussions with a great number of people. I would like to thank Guy Benveniste, Martin Carnoy, Robert Cole, Paul Dourish, Claude Fischer, Blanca Gordo, Bronwyn Hall, Marty Hearst, Nalini Kotamraju, Martin Kenney, Jennifer Kuan, Andrew Leonard, Aaron Marcus, David Mowery, Bonnie Nardi, Richard Nelson, Christos Papandimitriou, Walter Powell, Laurence Prusak, Richard Rosenbloom, Minna and Nir Ruckenstein, Annina Ruottu, Warren Sack, Pam Samuelson, Annalee Saxenian, Claus Otto Scharmer, Susan Stucky, Nancy Van House, Hal Varian, Bill Verplank, Georg von Krogh, Jack Whalen, Matthew Zook, John Zysman, and my friend and colleague from Nokia Research Center, Jukka-Pekka Salmenkaita.

Ikujiro Nonaka provided me with the opportunity to present an early version of chapter 7 at the Berkeley Knowledge Forum, and I have gained much from discussions with him in various places around the globe. Paul Duguid read carefully an early version of the manuscript and made many extremely useful, inspiring, and knowledgeable comments. John Seely Brown provided intellectual support and encouragement at various phases of the project.

Many of the ideas in this book were discussed among the members of the Menlo Circle, originally organized around the Stanford KNEXUS program. I would like to thank in particular Liisa Välikangas, George Campbell, Syed Shariq, Eilif Trondsen, Peter Coughlan, Renee Chin, Mahnoush Haririfar, and Helga Wild.

The Institute for the Future kindly invited me to join their Global Innovation Outlook program, which provided me with the opportunity to study the Silicon Valley innovation culture and get feedback on my work-in-progress. I would like to thank in particular Marina Gorbis, Patric Carlsson, Rod Falcon, Paul Saffo, Jan English-Lueck, and Magnus Karlsson for useful discussions.

I am also thankful for all the intensive discussions among the members of Sitra's innovation research program, and in particular for the comments provided by Gerd Schienstock, Timo Hämäläinen, J.-C. Spender, Reijo Miettinen, Aija Leiponen, Peter McGrory, Tanja Kotro, Rogers Hollingsworth, and Antti Hautamäki. In Europe, Japan, and the US, I have also benefited from discussions with Aleksi Aaltonen, Marko Ahtisaari, Jean-Claude Burgelman, Kathy Curley, Kirsten Foot, Kinji Gonda, Sara Heinämaa, Jeremy Hunsinger, Juha Huuskonen, Yrjö Engeström, Jyri Engeström, Sirkka Jarvenpaa, Petri Kasper, Kari Kuutti, Dorothy Leonard, Tarmo Lemola, Irma Levomäki, Tuomas Lukka, Teija Löytönen, Ian Miles, Hajime Oniki, Mika Pantzar, Matthew Ratto, Tuomas Toivonen, Linus Torvalds, Ryoko Toyama, Paavo Tuomi, and Jaakko Virkkunen.

Indeed, without interaction and networks no innovation could happen and no knowledge could be created.

I.T.
23 February 2002
Helsinki

CONTENTS

LIST OF FIGURES

LIST OF TABLES

1

Introduction

According to user surveys, the Linux operating system is rated as the best operating system available. It is considered to be more reliable than its main competitors. Its functionality is claimed to be better, and according to many experts, new releases of Linux implement innovative ideas faster than its competitors. In other words, it is argued that Linux development creates complex new technology better and faster than the biggest firms in the software industry.[1]

Yet, Linux also seems to break many conventional assumptions that underlie research on innovation and technological change. Linux is developed by an informal self-organizing social community. There is no well-defined market or hierarchy associated with it. Most of Linux development occurs without economic transactions. Instead of getting paid for their efforts, the developers often spend a lot of money and effort to be able to contribute to the advancement of the development project.

The open source development model, which underlies Linux, has attracted increasing attention in recent years. Today, Linux is considered to be a serious threat to Microsoft's market dominance in operating systems. More generally, open source development projects have in recent years had a major impact in software and internet-based industries. For example, almost 60 per cent of Internet connected Web servers were open source Apache servers in October 2000. As can be seen from Fig. 1.1, the second most popular Microsoft servers were about one third as popular with 20 per cent. Although Microsoft has gained market share with its Internet Information Server, at the end of 2001 about 63 per cent of active web sites were running Apache. The most common operating system in the web server machines was Linux.[2] Some open source projects, such as Sendmail, Perl, and Emacs, have achieved large user bases, making it difficult for commercial enterprises to enter the market.

Linux has been developed in the open source mode to a large extent because the Internet itself was to a large extent developed in this same mode. The collaborative

---

[1] http://www.uk.linux.org/LxReport.html.
[2] Source: Netcraft, http://www.netcraft.com/survey/. For a discussion on server market shares, see Netcraft and Peeling and Satchell (Peeling and Satchell, 2001).

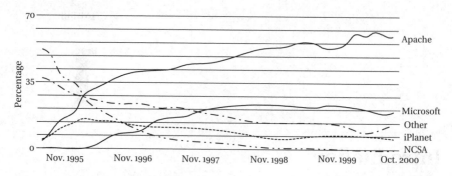

Fig. 1.1. WWW servers connected to the Internet
Source: Netcraft, http://www.netcraft.com/survey/.

and participatory development model gained visibility in the mid-1960s, when the early users of time-shared computers realized that collaboration often produced unexpected benefits. The predecessor of the modern Internet, ARPANET, was created in this mode, and many critical contributions, such as Internet email, Usenet news, and the World Wide Web emerged as a result of open collaboration. The Internet Engineering Task Force, which defines standards for the Internet, has also used an open source approach since its formation in 1986 (Bradner, 1999).

Several commercial software firms have recently tried to adopt aspects of the open source model. For example, Netscape announced in 1998 that it would distribute the source code of Netscape Communicator with open source licence. IBM decided to use the open source Apache server as the core of its Web server offers. Red Hat, SuSE, Caldera, and other new economy firms, in turn, make their business on packaging Linux distributions and by producing added value for Linux users. Sun Microsystems has used a version of the open source model to support development of its Java and Jini platforms. After launching an attack on Linux in 2001, Microsoft declared that it will have its own Shared Source Philosophy, which was aimed at making open source development possible without losing intellectual property rights. In all these cases, business firms are experimenting with ways to benefit from innovation that occurs in the open source communities. Instead of traditional economic competition, such initiatives rely on symbiotic relationships, and on the willingness of developer communities to collaborate.

In much of the innovation literature, innovation is defined as something that has economic impact. Linux and other open source initiatives show that this definition is problematic and possibly misleading in important practical cases. For example, during its history, most Linux development has occurred independently of direct economic concerns. It would be tempting to argue that Linux development is different from 'economic activity' and something that, strictly speaking, should not be called innovation. Indeed, in its early history Linux development was not in any obvious way associated with changes in production functions, market competition, or appropriation of economic investment and surplus. Yet, obviously Linux developers collectively produce new technology. If economy is about collective production, this is it.

Linux, therefore, is an interesting test case for economic theories of innovation and technology development. For example, the history of Linux allows one to question to what extent existing economic models of innovation and technological development capture phenomena that underlie collective production of new technologies.

In very practical terms, Linux is an economically important phenomenon. Indirectly, the success of many new businesses, venture capitalists, investment funds, and individual investors critically depends on the productive activities of the Linux community. Today, many corporations, governments, public sector organizations, and individual developers are starting to deploy Linux to cut costs, promote interoperability, and avoid lock-in to proprietary systems. Yet, when we consider the entire history of Linux, the economic impact seems to appear almost as an afterthought and as a side effect of a long period of technology creation. Linux, therefore, provides an interesting history of globally networked innovation, illustrating the substance that underlies the discussions on the 'new economy'. If the 'new economy' is about global Internet-enabled and software-driven production, this is it.

More generally, the history of Internet-related innovations enables us to discuss those social and cognitive phenomena that underlie technological change. By studying such innovations, we can open some black boxes of innovation theory, including such widely used concepts as learning, capability, utility, and consumption. By observing the development of the Internet, we can describe the microstructure of innovation, and transcend the boundary between invention and innovation.

Although such studies have obvious consequences for innovation research in general, Internet-related innovations are, however, also special. On the Internet the products of innovative activity are externalized as technological artefacts and documents that can be studied relatively easily. Never before has innovation and its results been recorded in such historical detail. On the net we live in dog years, but our memory is that of an elephant. There exists sufficient documentation so that we can—at least tentatively—describe some key principles that underlie the development of Internet related innovations. For a researcher on technological change, this is an exciting opportunity.

Internet related innovations are obviously important as the Internet has become a key technology in many areas of our everyday life. Below I will argue, however, that these innovations reveal important aspects of all innovative activity. Indeed, my key message is that the traditional models of innovation are often misleading, and that they will become increasingly misleading in the future. In practice, we have to move beyond abstract descriptions and ask what makes novelty meaningful. This leads to social and cognitive theories of innovation.

From a practical point of view, Internet related innovations also provide test cases for analysing product development models and proposals for organizing for innovation. For example, the extensive use of modern communication and collaboration technologies in Linux development highlights some aspects of technology development that were not easy to see in earlier studies on innovation. Although I will not explicitly discuss organizational or policy implications below, I believe that the following chapters highlight several points which have such implications.

Linux, open source projects, and Internet-related innovations may have developmental histories where collaboration and networking are more visible than in some earlier innovations. The open source model, however, obviously goes beyond software programming projects. As many commentators have observed, the process of science itself is very much based on peer-review, incremental development, non-economic motives, and geographically distributed collaboration. Indeed, traditional models of innovation often assumed that basic research generates ideas and technologies that are appropriated by entrepreneurs who turn them into products and money. The history of Linux and Internet-related innovations enables us to see how the boundaries between basic and applied research are being transformed. Indeed, I will argue below that the distinction between basic and applied research needs to be reconsidered.

From the very beginning, the Internet has been used to distribute work and its results. Division of labour is the foundation of all societies; the Internet, however, makes it possible in qualitatively new ways. A study on Internet-related innovations, therefore, has implications when we try to understand the ongoing social transformation towards the network society. To give just one example: when NASA ran its Clickworkers pilot where volunteer Internet users could mark craters on pictures of Mars, between December 2000 and June 2001 people marked over 1.9 million craters. Although each volunteer only marked a few craters, collectively their results were indistinguishable from those of a well-trained expert.[3] This example is interesting as it shows that trivial individual effort may lead to a high-quality collective outcome. In a very concise form it shows one way by which a new balance may emerge in the network society between increasing specialization and network-enabled participatory decision-making. Internet-related innovations, therefore, have relevance both when we try to understand how new technologies are developed but also when we try to understand how technological development and social change could be linked in the future.

History is always constructed from the perspective and for the purposes of the present. A useful history, however, provides opportunities for more than one interpretation. Historical description, therefore, has to be rich enough in detail and it has to give room for multiple voices. Yet, a balance has to be found between details and conciseness. Reality is always richer than any of its descriptions. I have tried to solve this problem by combining relatively general conceptual arguments with outlines of specific innovation histories and more detailed in-depth case studies. Some chapters make rather controversial theoretical claims without extensive empirical support for these claims. Subsequent chapters, hopefully, fill in some of the details.

The next chapter introduces some main concepts and assumptions that underlie the present work. In effect, it tries to set the reader in a position where the subsequent discussion can make sense. It points out that innovation is fundamentally about social change, and that innovations emerge and become articulated when they are taken into meaningful use in social practice. It argues that meaningful

---

[3] http://clickworkers.arc.nasa.gov/documents/crater-marking.pdf: 'Clickworkers results: crater marking activity', 3 July 2001.

use—as well as the meaning of technology itself—is grounded on social groups that can be called practice-related communities. As a result, innovation and technological change can be studied as phenomena that occur within an ecology of such communities. Construction of technology requires construction of meaning, and new technology is much more than improved functionality. Instead of the 'upstream' of the traditional linear model of innovation, we have to focus on the 'downstream' where social communication and change occurs. All innovation is social innovation. Innovation does not happen 'out there' in the world of objects, but in society and in minds. More particularly, it happens in the minds of the users, which are intrinsically integrated with the activities of the users. Those cultural and material resources that are available for the users, therefore, become key resources in the innovation process.

The third chapter is a quick first take on making these concepts more concrete. It illustrates the nature of innovation by outlining the history of the World Wide Web. It asks who invented the Web, what were the resources used in its invention, and what actually was invented in the process. Many of the details of this history are well known. Many accounts of the history of the World Wide Web, however, also show that some details of the story are often missed. These details become important when we try to understand innovations such as the World Wide Web.

The fourth chapter moves from recent history back in time, describing the early phases of the evolution of the Internet. More exactly, the focus is on that point of time when computer networking was only an idea. The chapter introduces the historical data that will be used in subsequent chapters. Although there now exists excellent histories of the Internet, such as those written by Abbate (1999) and Naughton (2000), it is necessary to provide enough historical detail to make the origins of the Internet understandable. In the process, I will also make some notes that hopefully complement existing histories in interesting ways. The chapter describes how electronic communication systems evolved and laid conceptual and material foundations for computer networks. It also introduces leading actors who played key roles in the early phases of computer networking.

The fifth chapter summarizes the early history of the Internet and describes the various technological frames that generated the basic innovations of computer networking. In other words, it puts history in the context of technology and innovation studies. It also discusses resource mobility in the early phases of the Internet development. One main claim in the book is that innovation occurs when social practice changes. The mobility of resources, therefore, is a key factor in enabling and constraining innovation.

The sixth chapter returns to the topic of communities. It discusses several alternative theoretical traditions that have described the social basis of meaning, knowing, and knowledge creation. It starts by introducing the concept of thought community that was originally introduced by Ludwik Fleck (1979) in the 1930s. Fleck's historical study described many of those social processes that underlie the emergence of new scientific knowledge and new technologies. The chapter further discusses Bakhtin's speech genres, cultural-historical activity theory, social learning in communities of practice, and the concept of *ba*. Ikujiro Nonaka and his colleagues have argued that innovation and knowledge creation occur in knowledge creation spaces, or *ba*s. The chapter discusses the nature of *ba*s, and links this concept back to its origins in the epistemological theory of Kitaro Nishida and the Kyoto School. The sixth chapter, therefore, introduces a set of alternative theoretical views that can be used to understand the cognitive and social basis of innovation.

One of the main arguments below will be that innovation can properly be understood only by studying the social basis of innovation. The heroic individual innovator is not a good model when we try to understand the evolution and development of technology. If knowledge and the meaning of technology is grounded in communities that reproduce existing social practice, as this book argues, it may seem, however, that innovation is a contradiction in terms. How is it possible that new social practices emerge when communities more or less by definition reproduce their current practices? How do we break technological frames and how are new technological frames created? Chapter 7 argues that there are two distinctive ways that new communities and new technological practices can emerge. One is based on increasing specialization, and the other on combination of existing resources. In other words, there exist two qualitatively different dynamics of innovation, and their analysis requires two different theoretical approaches. As a result of these two different modes of socio-technical evolution, the concept of *ba* can therefore be redefined. The chapter links the concept of *ba* to the sociocultural basis of knowledge, and proposes a new interpretation of Nonaka's knowledge creation model.

Using these theoretical concepts, Chapter 8 then returns to the history of the Internet. It briefly discusses email as an example of combinatorial innovation, and describes the evolution of the social structure that provided the basis for the creation of ARPANET and the Internet. It shows, for example, that both resource combination and evolution of specialization have played important roles in the development of social structure of Internet-related innovation communities. The current Internet community is in many ways rooted to the Network Working Group, which started in 1968 as an informal group of computer students. Internet, itself, however, would not have been possible without a combination of resources that came from outside this nucleus or the Internet culture.

Chapter 9 picks up one aspect of this history, which is an interesting topic for both innovation studies and policy. This is the question of retrospection and attribution of authorship. If innovations are to an important part created by their users and the meaning of innovation is reconstructed from the present position, how should we read historical accounts that describe evolution of technology? And to whom should the credit go? Did Al Gore really invent the Internet? Or was he just doing what Rembrandt did: signing off works that, strictly speaking, were produced by others, but which could not have existed without him? Should Linus Torvalds get a patent on Linux? What, indeed, does intellectual property mean when technology development uses resources that are networked, cumulative, often unintended, and when adaptation of new technological opportunities depends on institutional change and competence development in the downstream? Should we reconsider the author, or is the confusion created by a wrong conceptualization of the products themselves? By analysing newspaper articles that have discussed the Internet during the last fifteen years, we show how the common understanding of 'the Internet' has evolved. As Chapter 9 shows, the heroes of innovation are mental reconstructions, but so is the technology itself.

Chapter 10, finally, returns to the case of Linux. It describes both social and technological evolution of Linux and its development community. For example, it shows how technological architecture and social structure co-evolve as technical problems are solved in the social domain and social problems are solved in the technical domain. By analysing in detail the evolution of the structure of Linux source code over a period of years, it shows how social control and coordination become embedded in a technological artefact. It also shows how social interaction can be 'translated' into resources by 'black-boxing' some of the underlying complexity behind technological interfaces. The chapter argues that one reason why the open source development model has been successful is that the social translation mechanisms it uses allow several communities to interface simultaneously to a common technological artefact. Moreover, the open source model guarantees that when software fails, it fails gracefully, at least in the social sense. In open source, black boxes have transparent and penetrable walls. The chapter also discusses the bug removal process in Linux and highlights some trade-offs that are needed to make distributed innovation and technology development effective.

The last chapter puts the open source model of technology development in a broader perspective, and discusses the cultural and value system that underlies open source. Indeed, it argues that a study on socio-cognitive basis of innovation leads to a new approach in economic theory, where the concept of value has to accommodate the idea that in innovation processes new meaning is created and new domains of social practice are generated. Such 'expansive' theory of economics may lead to new insights when we formulate and study technology and innovation policy. The chapter also points out that the networked mode of production that underlies open source may lead to new dynamics in socio-economic development as the social institutions that usually provide stability in socio-economic systems are constantly renegotiated in the network mode of development. The chapter also discusses the differences and similarities between the open source model and the Silicon Valley innovation system. The chapter finally points out some areas for further study, and ends with some concluding remarks.

Innovation as Multifocal Development of Social Practice

Popular accounts and histories on innovation often focus on inventors and inventions. The creative genius of an inventor is commonly viewed as a force that produces new technologies and reveals hidden laws of nature. Such popular accounts, therefore, tell us, for example, how James Watt invented the steam engine, how Thomas Edison developed electric lighting, and how Tim Berners-Lee created the World Wide Web.

Research on innovation has conceptually refined this model by separating innovation from invention. *Invention* has generally been understood as it was described in the popular accounts, as a process of creative insight and heroic efforts in problem solving. *Innovation*, in contrast, has been defined as a process that refines inventions and translates them into usable products.

This traditional view led to a linear model of innovation. According to this model, innovations are first invented and then developed, packaged, marketed, and, finally, taken into use. Following Schumpeter (1975), Usher (1954), and others, many authors defined the process of innovation as sequential phases of idea generation, invention, research and development, application, and diffusion. Many product development and innovation management models have been based on this linear model. Similarly, many theoretical models have been developed to describe and predict the adoption and diffusion of new products generated in this process.

Since the 1970s it has often been noted that the linear model is too simplified (e.g. Kelly, Kranzberg *et al.*, 1978; Kline and Rosenberg, 1986; Padmore, Schuetze, and Gibson, 1998). In practice, innovations emerge in a complex iterative process where communication, learning, and social interaction play important roles. Allen (1977; Allen and Cohen, 1969) and others observed that communication and flow of knowledge is critical in the innovation process. Rogers (1995), in turn, noted that communication among users is necessary for the diffusion of innovations. Von Hippel (1976; 1988) emphasized that users often play an important role in the process of innovation by modifying and improving products. Cohen and Levinthal (1989; 1990) argued that adoption of new innovations requires learning and development of competences by the potential adopters, whereas Nonaka (1988),

Dougherty (1992), and others (Brown and Eisenhardt, 1995) noted that internalization of customer and market knowledge is critical for successful product creation.

Today product development processes are often iterative and technologies are refined while products are designed. Product development requires multidisciplinary integration and creation of knowledge. Kodama (1995) argued that the conventional product development 'pipeline' is increasingly being replaced by dynamic 'demand articulation' where product concepts are created for non-existent virtual markets. In demand articulation, potential user needs are integrated into a product concept, and the emerging product concept, in turn, is decomposed into development agendas for its individual component technologies. Many product developers now use 'focus groups', study 'lead users' (Griffin, 1997; Cooper and Kleinschmidt, 1991; Urban and Von Hippel, 1988), use active exploration and experimentation (Lynn, Morone, and Paulson, 1997; Thomke, Von Hippel, and Franke, 1998), and create alliance networks to improve new product designs (Doz and Hamel, 1997).

Although innovation research has made impressive progress, it still often relies on a basic assumption that also directed the traditional view. The conventional view assumed that the product of the invention process has well-defined characteristics. For example, the patent system was designed on the assumption that the functionality of each invention could be described in detail, so that its novelty could be unambiguously decided. It was also assumed that invention had a well-defined author—the inventor—and a well-defined moment of birth.[1]

In the conventional view both the inventor and the invention were unproblematic and easy to define. The moment of invention created simultaneously both the inventor and the invention. Although it was well understood that the primary insight often required development before an invention was articulated as a working prototype and could be produced, the exact details of the process of invention were often considered to be irrelevant.

Economists opened the black box of technology when they realized that innovation is a driver for economic growth. At the same time, however, invention was put in its own opaque box, veiled under impenetrable layers of creativity and insight. As a result, technological development was conceptualized as consisting of two qualitatively different phases: invention and its subsequent development into a product.

Science and technology studies, however, provide ample historical evidence that this fundamental assumption is not valid in general. New technologies do not come into the world ready-made. Instead, they are actively interpreted and appropriated by existing actors, in the context of their existing practices. A single technological artefact can have multiple uses, and new uses may be invented for old artefacts. Often a product is used in unanticipated ways, and perhaps no one uses it the way its designers expected it to be used.

Where, then, can we find the author of an innovation? When, exactly, is a new innovation born?

---

[1] When patent systems were increasingly deployed in the 19th century, these assumptions were, however, often contested, cf. Machlup and Penrose (1950).

One way to see the limits of the conventional view is simply to turn it around. Instead of a heroic inventor we can focus on a heroic user. The traditional view assumed that invention happens when a new concrete artefact or mental insight is created. The alternative view starts from a different assumption. Innovation happens when social practice changes. If new technology is not used by anyone, it may be a promising idea but, strictly speaking, it is not technology. Similarly, if new knowledge has no impact on anyone's way of doing things—in other words, if it doesn't make any difference—it is not knowledge. Only when the way things are done changes, an innovation emerges. Therefore we can say that invention occurs only when social practice changes.

This view is a useful starting point and it is compatible with historical evidence. It allows us to rethink some common assumptions that have become so central in the traditional view that they have become quite invisible. Careful historical study of innovations also shows that there has always been a great abundance of ideas and visions, only a few of which ever change everyday life and social practice.

Let us, then, at least for a while, assume that this user-centered model is useful and provides new insights on the process of innovation. If we give up the idea of technological innovations as something fixed, we can note that technologies and technical products have 'interpretative flexibility', to use a term proposed by Bijker (1987). Different user groups and stakeholders impute different meanings to a given technological artefact. A given technological artefact can play several different roles in different social practices. Instead of being a well-defined 'objective' artefact, with characteristics that could be described without reference to social practice, the artefact in question has many, and possibly incompatible, articulations. These 'meaningful products' may develop independently of each other, and one technological artefact can embed several meaningful products simultaneously.

If we adopt this user- and practice-centered model of innovation, it is easy to see that innovation has many agents and that the process of innovation is distributed in time, space, and across groups that use technology for different purposes. The traditional model of innovation focused on a very special case of innovation. This was the case where the user was well-defined, predictable, and whose needs could be taken for granted. As will become clear below, it never accurately described how innovation happens.

2.1 PUTTING THE USER IN FOCUS

By defining innovation as something that generates and facilitates change in social practice, we put the user in a central place in the process of innovation. In a very fundamental sense, it is the user who invents the product.

For example, for many decades after the telephone was invented, it was marketed mainly for business use in the US. When the telephone was not used for business transactions, it was often understood as a broadcast medium. Telephone entrepreneurs tried to use the telephone to broadcast news, concerts, church services, weather reports, and stores' sales announcements. The telephone was also expected to be used for voting campaigns, long-distance Christian Science healing, and to broadcast lullabies to put babies to sleep (Fischer, 1992: 66). Interactive social use of telephone was neglected for a long time by the inventors and the industry. Social conversations and 'visiting' over the telephone were not uses that telephone was supposed to serve, and industry sometimes resisted such use. As Claude Fischer notes:

> The story of how and why the telephone industry discovered sociability provides a few lessons in the nature of technological diffusion. It suggests that the promoters of a technology do not necessarily know or decide its final uses; that they seek problems or needs for which their technology is the answer, but that consumers themselves develop new uses and ultimately decide which will predominate. The story suggests that in promoting a technology, vendors are constrained not only by its technical and economic attributes but also by an interpretation of its uses that is shaped by its and their histories, a cultural constraint that can persist over many years.   (Fischer, 1992: 85)

The telephone plays a very different role in the different communication cultures of Japan, China, the US, Spain, Finland, or Bangladesh. In a very fundamental and practical sense, the telephone is a very different thing in these different cultures. Moreover, although a technological artefact may remain similar in shape and functionality, it is constantly created by its users. To give a very simple and concrete example, much of the revenue and most of the profits of telecom operators in Europe originate today from SMS text messages. When this technology was defined as a part of the GSM standard, no one imagined the various ways the users of this technology would appropriate it.[2]

Innovation, therefore, is not generated only by scientists or engineers, and often they are not critical sources of innovation. In many cases, we can take the availability of science and engineering for granted. In many ways, the modern world is full of ideas and new technologies are rarely, strictly speaking, new. The traditional model of a heroic inventor is therefore losing some of its obviousness and descriptive value. The emergence of new innovations depends to a large extent on resources that are available for the potential users, as well as on constraints that limit change in their current practices. Therefore, to understand innovation, we need to understand technologies in use. How, indeed, does technology become part of social practice?

2.2 USE AS MEANINGFUL PRACTICE

The traditional model of technological innovation was based on the idea that the inventor and the invention are unproblematic. According to this view, a typical invention is a well-defined artefact with well-defined characteristics. Consequently, new uses are new ways of using this given artefact. In this view, the telephone, for example, remains the same even when new ways are found to use the phone.

---

[2] SMS was mainly intended to be used to notify phone users that they had voice messages waiting.

Use of technology, however, is not something that we can understand as a specific use of a given technology. To talk about something as technology means that we already assume some uses. The concept of technology doesn't exist without an implicit model of use. Technological objects are not something that we can discover from nature: they exist as material artefacts that embed uses. Technological artefacts are artefacts full of meaning. If this meaning is taken away, we are not left with the 'objective' object without subjective interpretations: instead, we are left with a pile of undifferentiated matter.

This is sometimes difficult to realize. We rarely meet technological objects without some interpretation of their meaning. Even when we have no clue what the object is supposed to do, we still normally assume that someone knows. Most current technologies come 'packaged' with standard ways of using the technology, and in everyday conversation we take these standard uses for granted.

For example, it is quite difficult for us to talk about a telephone without interpreting it as enabler of common everyday practices in which the telephone plays a central role. When it is used as a hammer or a weight in a fishing net, for example, we may find this amusing or exceptional. Such exceptional uses, however, imply that there is a normal way of using the phone, and a corresponding standard interpretation of the meaning of the thing. In some circumstances a telephone may be a perfect hammer; most of the time, however, we are not supposed to use it as such. The reasons are complex and fundamentally social: for example, a telephone may have economically more efficient uses in a given culture. There may be cheaper hammers lying around. A telephone embeds a complex system of economic and social relationships and if we break the phone by hitting a nail with it, these relationships may break as well.

Technology enters our life as a way to conduct meaningful social practice. Therefore technology does not exist in a 'pure objective form' outside the context of social practice. Technology always exists as technology-in-use, and it is, in general, impossible to find a stable core use which would well define the nature of a technological artefact.[3] There is no hard core of technology that would provide a fixed foundation for different variations of use; instead, there are multiple ways a given technology can be appropriated by different actors, and different ways these actors can integrate technological products in their everyday life.

Technology-in-use refers to meaningful use of technology. Meaningful use, in turn, is rooted in social practice. In social life, completely idiosyncratic and unique events make no sense, and they appear as random noise in the social sphere of

---

[3] Material artefacts, of course, have some uses that are more 'natural' than others. According to Gibson (1950), the world presents itself to us as 'affordances'. For example, a ladder affords ascent or descent, and a chair affords sitting. We, therefore, see ladders and chairs as such, instead of seeing meaningless objects, which only after information processing become infused with meaning (Tuomi, 1999: 115). More generally, affordances characterize the possible uses of things. For example, it is difficult to push with a rope. A rope affords pulling. But, although material objects afford some things and do not afford others, their meaning is not fixed by a specific given use. In some circumstances it is even possible to push with a rope. If the pushing, however, becomes common enough, we find a new name for ropes that are used for pushing.

interaction. Social practice therefore grounds collective meaning. Meaningful use of technology, consequently, is always inherently social and related to social practices.

The traditional view on innovation was based on the idea that innovation is well represented by the material object that embeds the invention. A prototypical innovation, therefore, was something like a steam engine or a light bulb. The user-centric view on innovation leads to a very different prototypical innovation. Instead of material it focuses on mental. In the user-centric view, a prototypical innovation is more like a word or a concept. A word acquires its meaning through the ways it is used in communication. The same word can be used in many ways and it can play different roles in different types of conversations.

Innovation is, therefore, as much about creating new meanings as it is about creating novel material artefacts. Or—more exactly—it is *more* about creating meanings than it is about creating artefacts. A steam engine remains a pile of metal without someone making sense of it in the context of ongoing social practice.[4] Artefacts enter social practice as meaningful objects and we have no way of talking about material objects except as meaningful objects. Whereas the traditional view saw technological innovation as something that generated functional objects, the user-centric view sees these 'objects' as carriers of social practice and as artefacts that embed theories of meaningful use.

Such theories of meaningful use are open for reinterpretation and for new applications. Even though a designer of a new product may have a theory of use, there is no guarantee that this theory works in practice. Moreover, a given technical artefact can be understood through different theoretical and conceptual frameworks. These frameworks can evolve in various directions, long after the design of the artefact is frozen.

One relevant actor in all this, of course, is the producer of a technological product. In the traditional industrial mode of production, the producer was a manufacturer. The product naturally played an important role in the manufacturer's practices and often was the focus of activity. But even in the most prototypical examples of mass production, such as Ford's Model-T, the producer had only a limited control of the ways the product was integrated into the everyday life practices of car users. The meaning of a car was not invented by Henry Ford and his engineers. Instead, it was created by the users. In a very important sense the users produced the car, and without their active production, Model-T would have remained a working prototype. So, while the practices of the manufacturer are important when we try to understand the emergence of an innovation, they do not determine the evolution of the innovation. The 'essence' of a specific technological artefact cannot be found only by asking what its manufacturer believes the artefact to be. Technical specifications of a telephone, for example, tell very little about the uses for which telephones are put in different cultures and contexts.

---

[4] Of course, already the idea of 'pile of metal' itself assumes that there are practices where metals are differentiated from formless matter and non-metals. A 'steam engine' becomes a steam engine in relation to uses where it is different from just a pile of metal.

2.3 PRODUCTION AS AN END

One way to view the role of manufacturer is to see it as one user of a technological product among others. The manufacturer tries to make money with the product. Although many users may also try to make money with the technology—for example, by cutting costs in their production—the manufacturer has a special intent because this is essentially the only motive for the manufacturer. Whereas for the manufacturer the technological product is an end in itself, for the other users it is a means of getting something done.

This manufacturer's perspective generates the modern economic view of the world and, on the other hand, only makes sense within it. As innovation research often focuses on the economic aspects of innovation, it is important to understand the character and limitations of this economic view.

The economic view of the world is a very special view. Indeed, it is a special case of the utilitarian view that underlies also the modern concept of technology. In the utilitarian view, the meaning of objects is their use value. In a consistent utilitarian world, there are no ends but only means.

Such a world without ends risks becoming a world without meaning. As a reaction to the emerging modern utilitarian views, Kant argued that to make values, ethics, and politics possible we have to break the infinite regression of means in the utilitarian system. In Kant's philosophy, the world is an instrument for our interests and needs but we have to regard human beings as special cases, and treat them as ends in themselves, never only as means for other ends.

Classical economic theory solved the problem of infinite regression of means in another, ingenious, way. It assumed that meaning is fully reflected in prices. Instead of grounding value to something outside the economic system, value became one aspect of the system itself. The infinite chain of means–ends relations was connected by profit, which became the end of all means.

As a result, the means–ends system of the economic view of the world is a closed system. Use value is conceptually and without a residue linked to exchange value. Structurally, the economic world parallels the Newtonian closed universe, where novelty is a contradiction in terms. Scientifically, it all looked very nice, two centuries ago.

The economic view is important in the modern world and many actors operate under its conceptual and practical constraints.[5] For economic actors, the economic system is a perpetuum mobile, where quality can only appear in terms of quantity

---

[5] For example, Craig Mundie (2001), Senior Vice President of Microsoft, argued in May 2001 that the software industry needs intellectual property rights because that is the only way profits can be reinvested in research and development, and that it is only by such reinvestment that successful technology corporations are able to grow and continue to provide societal benefits. Mundie, therefore, adopts the later Schumpeterian (1975) model, where corporate R&D units become central in the production of new knowledge and innovation. According to Schumpeter himself, this leads to the growth of the size of firms. Schumpeter, however, also noted that the underlying logic eventually leads to the collapse of capitalism.

and where economic growth is the mother of all ends. It is, however, also important to understand the particular limitations of this view. Perhaps the most fundamental assumption that is built into the structure of economic explanations is that the meaning of products and production can be understood without reference to the social and material reality. In a conceptually coherent and completely closed self-referential system of means–ends, there can be no loose ends. As the economic view implies in this sense a perfect utilitarian world, it is conceptually blind to anything that does not operate within a utilitarian logic. 'Values' therefore remain inherently external to the economic world and make no sense within this world. The meaning of economic life is profit. All other meanings are unlinked from the system, irrelevant for it, and fundamentally meaningless.[6]

It is always possible to introduce new theoretical concepts that expand any closed theoretical system so that any particular phenomenon can be explained within the system. The economic concepts of utility and preference, indeed, try to do something like this, and link the economic system back to the world. The reasonableness of such extensions, however, is also an empirical question. At some point we may end up with a complex Ptolemaic system and may ask whether new conceptual starting points could make our explanations more powerful and compact.

Turning back to Kant, for example, we might ask whether there are non-utilitarian ends. More fundamentally, we can rethink what makes products and production meaningful. Such a quest for 'meaningfulness' of technology, products, and innovation may at first look remote from the concerns of entrepreneurs, managers, and modern policymakers. There are, however, many ways to make profit. In the future good business may require understanding how meaning is produced.

2.4 INVESTMENT AND INVENTION OF MEANING

It is useful to illustrate these concepts by looking for innovations that are all around us, yet rarely discussed in innovation literature. During the last two decades, the active role of users has been one central theme in research on fashion. Fashion is an interesting reference point to studies on innovation as fashion is novelty without technological function. In other words, whereas we often assume that new technology is adopted because it in some sense 'works better' than existing technology, fashion has no obvious link between novelty and function.

Fashion is innovation without progress but also obviously social, and full of meaning. According to Grant McCracken (1988), the meaning of material goods has three loci, and there are several different ways meaning can be transferred from one locus to another. The original location of meaning is the 'culturally constituted

---

[6] The inherent destruction of meaning in the utilitarian world was discussed already by Nietzsche. Arendt (1998) has provided an insightful account of the conceptual structure and history of the modern economic world-view in her classic work, *The Human Condition*, originally published in 1958.

world':

This world has been constituted by culture in two ways. Culture is the 'lens' through which all phenomena are seen. It determines how these phenomena will be apprehended and assimilated. Second, culture is the 'blueprint' of human activity. It determines the co-ordinates of social action and productive activity, specifying the behaviors and objects that issue from both. As a lens, culture determines how the world is seen. As a blueprint, it determines how the world will be fashioned by human effort. In short, culture constitutes the world by supplying it with meaning.   (McCracken, 1988: 72–3)

According to McCracken, this meaning can be characterized in terms of two concepts: cultural categories and cultural principles. Cultural categories represent the basic distinctions with which a culture divides up the phenomenal world. These categories include categories of time, such as years, centuries, leisure time and work time, and sacred and profane time; and categories of space, flora, fauna, social classes, status, gender, age, and occupation, for example. Cultural principles, in turn, provide the principles that allow us to evaluate, rank, construe, distinguish, and interrelate phenomena in the world. Together, these categories and principles create a system of distinctions that organizes the world.[7] Culture, therefore, constitutes a world by investing it with its own particular meanings.

It is thus that each culture establishes its own special vision of the world and thus that it renders the understandings and rules appropriate to one cultural context preposterously inappropriate in the next. Culture makes itself a privileged set of terms within which virtually nothing appears alien or unintelligible to the individual and outside of which there is no order, no system, no safe assumption, no ready comprehension.   (McCracken, 1988: 73)

Cultural categories cannot be seen as such in the world. Instead, they remain invisible, at the same time providing a structure of distinctions that is continuously reproduced in social life. According to McCracken, consumer goods have an important role in materializing and substantiating these categories. Through goods the meanings that organize the world are made visible.

Cultural categories can be embedded in artefacts through the systems of advertising and fashion. Advertising brings a consumer good and a representation of the culturally constituted world together, so that the known properties of the world can become resident in the unknown properties of the consumer good (McCracken, 1988: 77). Similarly, the fashion system invests and divests of meaning in goods.

McCracken noted that the fashion world works in three distinct ways in transferring meaning to goods. One is related to advertising. The fashion system takes new styles and associates them with existing cultural categories. The fashion system, however, can also invent new cultural meanings:

This invention is undertaken by 'opinion leaders' who help shape and refine existing cultural meaning, encouraging the reform of cultural categories and principles. These are 'distant'

---

[7] Research on categorization has, of course, been a key theme in cognitive science. Gardner (1987) and Lakoff (1987) provide good introductions to the history of categorization research. More sociological studies on categorization and its consequences include Bowker and Star (1999), and Foucault (e.g. 1970).

opinion leaders: individuals who by virtue of birth, beauty, celebrity, or accomplishment, are held in high esteem.   (McCracken, 1988: 80)

The third way the fashion system works is when it reforms cultural categories and values that define cultural principles. The groups that are responsible for radical reform of cultural categories usually exist at the margin of society. These groups may adopt cultural categories and principles that differ fundamentally from the traditional ones. For example, hippies and punks redefined the categories of age and status, and gays redefined the category of gender. At the same time these groups became 'meaning suppliers' for the mainstream culture.

In the fashion system, journalists, social observers, and market analysts act as gatekeepers, filtering aesthetic, social, and cultural innovations, judging some as important and some as trivial. 'It is their responsibility to observe, as best as they can, the whirling mass of innovation and decide what is fad and what is fashion, what is ephemeral and what will endure' (McCracken, 1988: 82). After making their selections, they engage in a process of dissemination with which they make their choices known. These cultural innovations are then invested in products by designers who create fashion products for mass consumption.

By consuming the products, the meaning can transfer to the consumer. McCracken proposed that there are four ways this transfer can occur. First, there exist various possession rituals. Consumers may clean, discuss, compare, reflect, show off, and, for example, photograph their new possessions. While all these activities have an overt functionality, they also enable the consumer to claim the possession as his or her own. In addition to such possession rituals, there exist also exchange rituals, where a consumer chooses, purchases, and presents goods as gifts. The gift giver, for example, may choose a gift because it possesses meaningful properties that the giver would like to see transferred to the gift-taker.

A third type of meaning transfer ritual is grooming. Some meaning is in its nature perishable and has to be constantly recreated. The purpose of grooming is to 'take the special pains necessary to insure that special and perishable properties resident in certain clothes, certain hair styles, certain looks, are, as it were, "coaxed" out of their resident goods and made to live, however briefly and precariously, in the individual consumer' (McCracken, 1988: 86). The object of grooming can also be the good itself. For example, cars, computer games, and tamagotchi may become supercharged with meaning when they are groomed. This meaning, in turn, can be transferred to the person who grooms the car, game, or tamagotch, redefining the person as a proud owner of the consumer item in question.

When meaning becomes invested in a material artefact it may become part of the personality of the possessor. Divestment rituals may therefore be needed to separate the object from its consumer. For example, when someone buys an object that has previously been part of the personality of someone else, divestment rituals may erase the meaning associated with the previous owner. According to McCracken, the cleaning and redecorating of a newly purchased home may be seen as such an effort.

McCracken noted that in North American culture, cultural categories and principles are exceptionally indeterminate. People can define and choose their 'age', 'gender', 'social class', and other category memberships in ways that are often strictly specified and policed in other cultures (McCracken, 1988: 74). Indeed, much time and energy goes to managing such memberships in a culture where they are not given.

In a modern world, cultural diversity supplies huge amounts of potential meanings for everyday use. The resulting instability of cultural categories creates a paradoxical situation. Individuals have to consume mass produced goods to differentiate themselves. This is not only an economic decision, or an attempt to minimize the cost of constructing one's identity. For differentiation to make sense, it has to be based on commonly known cultural distinctions. Mass consumption, in other words, requires mass media. Cultural contingency leads therefore to a situation where cultural categories tend to be highly visible. Such a world looks open to innovation and individual expression, but it also constrains novelty and expression. To rephrase Henry Ford in the age of electronic highways: you can select any age, as long as it looks young.

As Thompson and Haytko (1997) noted, current fashion is also often seen as a common 'generalized other' against which differences can be made. For example, in the interviews conducted by Thomson and Haytko, some students actively tried to avoid becoming 'a statistic' and dressing just 'like the others'. The perceived individuating and transformative power of clothing, however, is ultimately contingent upon a belief that others will notice and care about one's appearance.[8]

Thompson and Haytko argued that clothing often has a metonymic role. Metonymy, used in its normal linguistic context, means that a part, association or a property stands for a whole, for example, as in the phrase 'Wall Street focuses on technology stocks', in 'Paris wears yellow this spring', or 'he drank the whole bottle'. Clothing and material objects have a similar metonymic function when they situate an individual as a member of a particular social sphere. A tie, jewellery, baggy jeans, a yarmulke, or a turban all may stand for a whole lifestyle and a complex system of values.

Simple reproduction of cultural categories provided by the fashion-system is, however, an antithesis for a fashion-conscious consumer. Instead, the highly promoted 'brands' and 'looks' are used in developing one's own 'style'. The culturally available resources are combined and adapted to create something new. Indeed, in the world of US university students, the capability to create coherent ensembles from a range of brands and styles is taken to signify a number of positive meanings such as creativity, organization, competence, and conscientiousness (Thompson and Haytko, 1997).

The contingent nature of fashion means that the only way its different expressions can acquire meaning is through active discourse where these meanings are negotiated. There is no such thing as a personal fashion, as clothing derives its symbolic capital from cultural categories and principles. But neither can fashion be

---

[8] In practice, the system of fashion works well because it secures its own foundation. Fashion requires that we construct our identities 'under the objectifying gaze' of imagined others. As a consequence, the attitude that underlies fashion guarantees that we don't need to know what others really think. Although identity is fundamentally social, the fashion system enables identity construction that is based on mediated representations of others. For the same reason, it is also possible to dress fashionably 'just for oneself' even when there are no others present.

universal. Distinctions become invisible if everyone wears a similar dress. Similarity, therefore, defines an interpretative community, which in turn defines what it considers to be similar and where it finds differences.

The culturally constituted world, as described by McCracken, is therefore not a homogeneous world. Although generic cultural categories provide the backdrop against which distinctions can be projected, individuals negotiate and recreate the meanings of these distinctions as members of interpretative communities. Sometimes the community may limit the way new interpretations can be invented. For example, a similar uniform or gown may signal that some cultural categories are excluded from the ongoing discourse.

A closer look at the ways fashion is 'consumed' therefore reveals that goods are actively produced by the consumers. Fashion items are used in a discourse which defines and comments cultural categories. Pure fashion lacks functionality and therefore its use is purely communicative. The consumer is never a passive sink of goods or an end point of a production chain. Instead, products that are consumed are used in the production of the user. Only if we assume that these uses of the 'consumer' are irrele-vant or uninteresting, can we forget the processes by which the 'consumer' makes a product meaningful.

In industrial products fashion is often regarded as irrelevant and the fashion industry usually regards functionality as irrelevant. In both cases, the adoption of new products, however, is based on the users' capability to make sense of the product and to integrate it into ongoing practice. This practice, in turn, defines a community that sustains and reproduces the practice in question. Research on fashion highlights the point that the resources needed to create new social practices and categories are socially distributed and produced, and that, for example, communication, advertisement, and the reputation of opinion leaders play an important role in their change.

In effect, we therefore redefine the producer. The traditional view on innovation assumed that a producer is either an inventor or an entrepreneur who produces a new innovative product and develops a market for it. A user-centric view on innovation, in contrast, sees the traditional inventor and the entrepreneurial innovator as users among other users. They have specific roles, competences, and motives but in that regard they do not fundamentally differ from other actors that collectively co-produce innovations as meaningful products. Innovations are produced through interaction between the different users, and innovation therefore cannot be localized within a single business firm or in the head of a single inventor. As the following chapters show, this has important implications for the evolution of technologies, as well as the ways innovative activity can be organized.

2.5 COMMUNITY AS THE LOCUS OF PRACTICE

The locus of innovation is a group of people who reproduce a specific social practice. Social practice does not exist in a vacuum, and it is not something that an individual can invent on her own. There are no more private social practices than there are private words. Language and practice are both inherently social. A new word can be created by an individual but it becomes a meaningful word only if it is taken into use in language. Similarly, innovations become innovations only when they start to play a role in meaningful social practice.

Social practice consists of reproduced forms of action. Technological artefacts often play an important role in the formation of social practice as they externalize aspects of practice and transform parts of it from the mental sphere to the concrete material world. Practices, therefore, exist as complex networks of tools, concepts, and expectations.

When we talk about a given social practice, we therefore assume that there is a recurrent form of activity that has some stability. Social practice structures and organizes social life, and provides a foundation for collective meaning processing. This foundation is not fixed but it provides a practical basis for interpreting the world. As practices comprise complex heterogeneous networks of artefacts, concepts, and ongoing social activity, reconfiguration and evolution of practice has many constraints.

Meaning is not something that can be grounded on individual decisions or cognitions about the world. But neither it is something that can be derived from some abstract structure of society. Instead, meaning is grounded on specific communities that produce and reproduce meaning and their unique ways of knowing the world. Meaning has its origins in collaborative practical activity and the community that reproduces specific meanings is the community that reproduces the related practices. The foundation and carrier of social meaning can therefore be called a community of practice.

There exist different proposals on how we should conceptualize such communities of practice. Some authors have focused on communities of identity and interpretation, others on communities of production, competence development, or communication and knowledge creation. These proposals are discussed in more detail in subsequent chapters. At this point, we may simply note that a community creates specific potential uses of technology. The 'user' of technology, therefore, is not an individual person but a member of a community with a practice that uses the technology in question. The individual user is engaged in the practices of the community and makes sense of technology in the context of these practices. When innovation changes these practices, new ways of doing things create new interpretations of the world. If innovation is technological, technology becomes integrated in social practice in new ways, and acquires new meaning.

As a user of a given technological product, an individual is a carrier of social practice. In other words, the user is conceptually more accurately described not as a person but as a practice.

This is an abstract conceptual point, but it is also a very important one. Innovation studies often adopt an individualistic and object-centric view. This subject–object dichotomy is deeply ingrained in our language and conceptual systems. When we try to describe alternatives for it, our concepts easily start to look imprecise and not well defined.[9] This is because concepts make sense only as parts of larger conceptual systems. Therefore it is impossible just to define a concept such as a 'user' in a new way, without simultaneously changing the relationships between many other concepts.[10]

For the purposes of the current work, it is not necessary to deal in any great detail with these conceptual issues. Instead of defining concepts we can describe concrete examples and historical cases of innovation. By illustrating the processes of innovation we can provide material that allows us to redefine the way we use existing concepts. It is, however, important to note that to the extent that the meaning of technology cannot be grounded on interpretations of individual actors, models of innovation cannot be based on individual users.

The strong theoretical claim that underlies the present work is that technology exists as technology-in-use, in a context of a specific practice. If we abstract away this social practice, we end up with an individual user or aggregate groups of similar users. At the same time, however, we also abstract away technology itself. The conceptual starting point for innovation studies, therefore, has to be at the level of social practice.

2.6 INTERPRETATIVE FLEXIBILITY AND ECOLOGY OF SOCIAL PRACTICES

The user-centric view on innovation means that the various stocks of meaning available to the different users provide the basis from which the innovation is articulated. Innovation, therefore, is not created just by using the resources available for the producer. In this sense, the innovation process is distributed among a number of stakeholders. Innovations are generated in the interaction between the various users and the artefact that embeds the innovation in a concrete form.

There are many ways of using a given technological product and there are several communities of practice that have their idiosyncratic views on the meaning of the product. The traditional view on innovation often noted only two of these: the producer and the consumer. In practice, there are many different 'consumers' who 'consume' the product in their own productive practices. If we follow a given artefact and register all the different communities where it is used, we therefore may

[9] As Shapin (1995) noted in his overview on the sociology of scientific knowledge, any discussion on the alternatives to the object-centric conceptualization of the world has to rely on object-centric concepts if it wants to be intelligible.

[10] When we talk about the user as a social practice, we therefore implicitly switch to a new view of the world where many concepts acquire new meaning. This is, more generally, the key characteristic of radical innovations. Radical innovation requires that we change some of the central concepts and practices in a given community and reorganize the system of activities and meanings in a discontinuous way.

find many such communities. In each community, the meaning of the artefact is different. Only rarely is there just a single community of users.

For example, word-processing software is an object of development and manu-facturing for people who work in the firm that makes the product. Within the firm there may be many different communities that deal with the product in different ways. For example, product developers may see it as software code, marketing people may see it as a solution to customer problems, and the finance department may see it as a source of costs and revenue. For store managers it is an object that needs to be shelved and sold. For technical support people it is a collection of doc-umented and non-documented features and bugs. For others, a word-processor can be a tool to write poems, business letters, and technical documents. What a word processor is depends on how you use it. How you use it depends on what prac-tices you are engaged in. A word processor, therefore, can be many different things in different contexts of use.

In a sense, these different uses may look trivial. A single product obviously has different functions for its manufacturer and customer, for example. The point here, however, is more radical. Strictly speaking, there is no 'single product'. Such a thing simply doesn't exist. This becomes important when we try to understand the evolu-tion and life cycles of technical innovations.

Although software products may seem exceptionally generic and multipurpose products, they are not fundamentally different from any other technological products. Wheels, forks, cars, light bulbs, telephones, computers, email, and voice recognition systems are all multipurpose products and have multiple communities of practice using them. Some technological products are, of course, difficult to adapt to different uses. For example, a nuclear power plant may produce heat and electricity, and there may be constraints on the ways the plant can be used for alter-native purposes. It may, for instance, be difficult to turn a nuclear plant into a play-ground for children. But although there are limits to the ways a given technological product can be used—and many practices exist where the product is not used—in general there are many user communities and many interpretations of the product.

A given technological product evolves in the context of these different interpreta-tions. Different stakeholders make different claims concerning the product. Some stakeholders may dominate this process and make some interpretations socially more legitimate and visible than others. Some uses and user communities may become dominant, others may become peripheral, and some users may remain marginal. In the course of evolution of a given technological product the centrality of different communities may change and latent uses, for example, may become dominant. Although the 'product' itself may remain similar in design, its meaning may change.

An innovative product, therefore, is about co-development of practices and a meaningful product that plays a role in those practices. Sometimes innovation can occur simply when the meaning of an existing technological artefact is reinter-preted and appropriated in a social practice where the artefact was not previ-ously used. Innovation does not necessarily require change in the design of the product.

The development of a given technological product is therefore a continuous process. The evolution of its functional characteristics reflects a process of social differentiation and negotiation of interests. When the design of the artefact changes, some of the tensions in the underlying social processes become concretely embedded in the product. At the same time, the artefact may 'freeze' some of this underlying social structure. Some technological designs provide ample opportunities for continuous innovation whereas some designs effectively implement a fixed view on the use of technology, thus limiting the opportunities for innovation.[11]

We should therefore understand innovation as a multifocal process of development where an ecology of communities develops new uses for existing technological artefacts, at the same time changing both characteristics of these technologies and their own practices. Some of these communities have, of course, a more prominent position in this process than others. Indeed, the traditional model assumed that the only relevant communities are the producer community and the primary user community. In many industrial products this assumption was a reasonable one. An iron plough, for example, may have its most relevant use in agriculture, and many of its uses are relatively easy to predict. Many modern products, however, have flexible uses and many user communities. The drivers for innovation cannot easily be found by looking at a single group of users or by searching the source of invention from the deep well-springs of individual creation. Innovation is a social phenomenon. It is generated in complex interactions between several communities, each with their own stocks of knowledge and meaning. Technological designs and social practices co-evolve. Therefore all innovation is fundamentally social innovation.

2.7 SOCIAL DRIVERS OF INNOVATION

As was noted above, in the multifocal and practice-centered innovation model, innovation occurs when social practice changes. Drivers for innovation can therefore often be found by looking for tensions and contradictions in existing social practice. Social practices form a complex network of interlinked practices and this network is continuously evolving. Technology addresses a need when it releases or reduces some of the tensions generated in this process.

---

[11] The problem of flexibility, of course, is becoming an increasingly important challenge as product life cycles decrease. Research on product development models has traditionally focused on process flexibility and product flexibility (Adler, 1988). Process flexibility means that manufacturing capability can be easily reorganized and product flexibility refers to the ability to create product variations. More recently, researchers have studied ways the product creation process itself can be made more flexible (e.g. McKee, 1992; Mullins and Sutherland, 1998; Bhattacharya, Krishnan, and Mahajan, 1998; Verganti, 1999). Bhattacharya et al., for example, discuss the problem of creating product definitions in highly dynamic environments where customers cannot easily articulate their needs and where these needs may rapidly change. New product development models, however, do not consider interpretative flexibility, or user meaning creation and innovation.

Entrepreneurs develop new technological products with the explicit purpose of addressing needs. An entrepreneur interprets the meaning of technology from the point of view of a potential user and designs a product that addresses the user's need. When the product addresses a need that articulates an important tension in the underlying network of practices, the product can be quickly adopted.

The entrepreneur may also address latent needs that have not yet been articulated as well-defined needs. In this case the entrepreneur invents both the need and the product that addresses the need. This is the process that Kodama (1995) called demand articulation.

Of course, there are limits within which needs can be invented. It is difficult or impossible to articulate needs that have no basis in the current social life. A successful entrepreneur is in this sense like a popular poet who puts in words what everyone was thinking but no one had said before. If a new product gives a compact and coherent expression to something that was not expressed before, the product can become an important element in the construction of social life.

Often such innovations, however, fail. The entrepreneur is rarely able to see all the constraints that underlie the forms of current practices and how they limit the ways practice can change. In other words, the entrepreneur may have a wrong model of the potential use. This happens easily because potential uses are potential: there is no simple way to observe such potential uses in their actual form. Although they can sometimes be 'simulated' and 'tested' they cannot be observed in their ecological context. As a result, the entrepreneur has limited possibilities to improve the imagined models of use.

Historical analysis of important innovations shows, however, that even when the entrepreneur has a wrong model of use, innovations often succeed. The entrepreneur produces a product for purposes that look relevant and important. Frequently, however, the product is used for different purposes, and these unintended uses may become key drivers in the evolution of the product. Sometimes the producer makes a good guess on how the users will understand the product, but often the guess goes wrong. Often the producer neglects some potential user communities entirely, and these forgotten communities may become main users of the product. Fundamentally, however, it is the users who either succeed or fail in making the product meaningful.

Entrepreneurial activity thus both addresses existing needs by reducing tensions in the system of social practices and creates possibilities for new forms of practice. In the latter case, we often talk about a 'solution looking for a problem'. In such cases, the entrepreneur may have a model of use that doesn't resonate with any potential user group. Many of the proposed early uses for telephone, television, and computer, for example, tried to articulate needs that didn't exist. People really were not that interested in listening to concerts using a telephone or maintaining recipes in computers—especially when they would have filled a kitchen.

In such cases, the locus of innovation moves away from the producer. Technological products become more like Rorschach ink-blots, and it is up to the users to figure out what they mean. The telephone can be appropriated for social conversations, television can become a medium for advertising, and the computer may become a communication machine. Technology creates interpretative flexibility and makes new forms of practice possible. Technology itself can therefore promote change. It can destabilize existing forms of practice and create contingency where it did not exist before. Technology is, as it were, thrown to the world for someone to pick it up and figure out what to do with it.

More often, however, technology is thrown to the world for specific expected uses. Also then it can be picked up and used in practices where it was not designed to be used. As will be shown below, this is quite a common event. In many ways the future of a given technological product is decided not by its designers but by the dynamics of an ecology of user communities where unexpected uses often become critical uses.

One way of seeing the dynamics of technological evolution is to view it as an integral element in a changing ecology of interacting social practices and communities that produce and reproduce these practices. Innovation often has its source in the needs generated when a network of practices produces tensions and searches for ways to reduce these tensions. One could call this the 'tectonic model' of innovative development. When social practices collide, technology is used to enable the formation of new shapes in the landscape of social practice.

2.8 INDIVIDUAL EXPLORATION

In addition to fundamentally social drivers of innovation there is also another source of innovation. In his well-known works on flow and creativity, Csikszentmihalyi (1990; 1996) argued that a generic feature of the human psyche is that humans feel happy when they successfully perform at the edge of their capabilities. At the individual level, humans have a tendency to do things that they find challenging, but which are within reasonable distance from their current level of competences. To feel happy, people are willing to take risks and perform at the limits of their competences. As a result, these competences develop and the domain of competence expands.

One interpretation of this observation is that humans are psychologically wired for extending behaviour beyond its current forms. People play with the limits and try to find ways to do things that were impossible before. Innovation, therefore, is not necessarily generated only by tensions in social systems; instead, it also has an independent counterpart in individual playfulness and the joy of exceeding the given limits of the possible. Indeed, individual creativity often drives change in social practices and also creates tensions in the process.

In practice, play, creativity, needs, and opportunities are closely linked. Existing designs are often improved because it is possible and fun, and because the improved practice and design provide happiness and aesthetic satisfaction (Pacey, 1999). Improvement, therefore, cannot always be reduced to better functionality. Sometimes an improved product just feels good.[12] The product can be meaningful

---

[12] The aesthetic dimension of innovation is particularly evident in art. The earliest use of the term 'invention' in music dates back to 1555. In 1720 Antonio Vivaldi used the term in a somewhat Schumpeterian way, titling his opus 8 'The Contest Between Harmony and Invention'. The most

by articulating something which cannot be formulated as sentences or functional descriptions.[13] An improved product, for example, can embed a view of the world which shows that the world is richer or different than we expected. The product can be 'true' and embed insights that make us happy. A true product fits easily with our life. When we see such a product, we may laugh or smile, or feel at home with it. Sometimes, however, the best we can say about it is: 'this is it'.[14]

Although individual meaning is fundamentally grounded on social meanings, which we learn through our involvement with culture and its social practices, individual creativity can also produce new interpretations of the world and its artefacts. Individuals can, for example, apply concepts in new contexts. Indeed, Schön (1963) argued that such 'displacement of concepts' is a key source of innovation. If existing results of individual creativity are appropriated in social life, creativity can transform into innovation and become social. At the same time, the generated new practices become the new basis for socially shared meanings that provide the foundation for further evolution of technologies, practices, and systems of social meaning processing.

When there exists an obvious meaningful use for a technology, we can talk about demand pull as a driver of technology development. When the need has to be articulated and invented by the user community, we often talk about technology push. But often innovation is also driven by playful tinkering with the limits of possibility, in a process where new possibilities and new spaces for social practice are created.

Sometimes, in other words, we do it just for fun.

2.9 SPACES OF NOVELTY

For sure, innovators and revolutionaries don't always have fun. Social practice, by definition, is recurrent and continuously reproduced. The heroic model of innovation is based on historical reality: innovators often become excommunicated, beheaded, or bankrupt. Social practice is inherently conservative. How, then, is change possible in social practice? If knowledge and meaning have their roots in

famous inventions, however, are J. S. Bach's two-part inventions that explicitly aim at teaching the student how music can be invented by thematic transformations. Bach's inventions are especially interesting as they show how familiarity, novelty, and aesthetic satisfaction can be integrated using relatively simple processes of innovation.

[13] Indeed, Csikszentmihalyi and Rochberg-Halton (1981) observed that everyday objects often become carriers of meanings that may have little to do with their functionality. Although the socio-cultural origin of meaning for functional artefacts is in productive social practice, their meaning cannot be reduced to productive social practice.

[14] All artefacts play multiple roles in social life, for example, by being used in production at the same time when they are used in the reproduction of social practices. Such an artefact makes sense in an ontology that always remains only partially articulated and which we are never able to completely describe (Polanyi and Prosch, 1975). In this ontology, the functionality of things is only one of many relationships between the entities in the world, and therefore only one component of their meaning.

existing social practices and technology is used in these practices, how is it ever possible to create new knowledge and new technologies? In a world filled with social practices, where can we find space for something new?

The obvious answer is that, although new knowledge and innovation emerge from the basis of existing knowledge and system of meanings, they do not emerge *as* social practices. Novelty starts small. If it leads to innovation, it expands from its origin and becomes institutionalized. This process of expansion and institutionalization is a key component in the emergence of a new innovation.

Innovation, however, is possible only because members of user communities can break the institutionalized forms of practice. As noted above, practice exists because it is regularly reproduced. The fidelity of this reproduction, however, is not perfect and there is variation in practice. More importantly, some rules can be extended, reinterpreted, or broken. Although society structures everyday practice, individual actors always deploy these structures for their own purposes. In this sense the teenager who browses the mall to find and express his or her individual fashion statement is right. Creativity can also be expressed by mixing and matching the different brands that hang on the racks.

Michel de Certeau (1988) argued that in modern society life is very much about improvising around existing forms of practice, and appropriating them for individual needs and idiosyncratic situations at hand. Although everyone necessarily has to live according to someone else's rules in modern society, the daily practice relies on tactics that divert existing resources for unintended uses. Everyday life is improvisation in the context of the current situation. Practical mastery is reflected in this capacity to appropriate given structures for one's own purposes:

People have to make do with what they have. In these combatants' stratagems, there is a certain art of placing one's blows, a pleasure in getting around the rules of a constraining space . . . Scapin and Figaro are only literary echoes of this art. Like the skill of a driver in the streets of Rome or Naples, there is a skill that has its connoisseurs and its esthetics exercised in any labyrinth of powers, a skill ceaselessly recreating opacities and ambiguities—spaces of darkness and trickery—in the universe of technocratic transparency, a skill that disappears into them and reappears again, taking no responsibility for the administration of the totality. (de Certeau, 1988: 18)

The individual actor, therefore, lives according to the constraints and resources provided by the social world around her, and manages her way in everyday situations as well as she can. Common forms of social practice are therefore not produced by replicating and reproducing them in any high fidelity; they emerge as a collective normal form of practice and as a standard against which exceptions can exist. The members of a community do not form a single collective mind. Instead, they have their idiosyncratic situations and history as the basis for their current action. Although social practice is in many ways institutionalized in technology, norms, and shared meaning, in many ways social order is an imaginary order. It is not always easy to think or act differently, but it is always possible.

Innovative use of given constraints and resources reflects the conflict between the necessities of practice and the dominant representations of appropriate practice.

According to de Certeau, in modern society there is little free space for individual producers. We are confined in systems of production where the only remaining task for us is to become consumers. Our creativity can be expressed only through the different ways we consume the products of the modern society. At the same time, however, we can create a space for our everyday practice within this system. An example of this can be found in what in France is called *la perruque*, 'the wig':

La perruque is the worker's own work disguised as work for his employer. It differs from pilfering in that nothing of material value is stolen. It differs from absenteeism in that the worker is officially on the job. *La perruque* may be as simple a matter as a secretary's writing a love letter on 'company time' or as complex as a cabinetmaker's 'borrowing' a lathe to make a piece of furniture for his living room. Under different names in different countries this phenomenon is becoming more and more general, even if managers penalize it or 'turn a blind eye' on it in order not to know about it. Accused of stealing or turning material to his own ends and using the machines for his own profit, the worker who indulges in *la perruque* actually diverts time (not goods, since he uses only scraps) from the factory for work that is free, creative, and precisely not directed toward profit. In the very place where the machine he must serve reigns supreme, he cunningly takes pleasure in finding a way to create gratuitous products whose sole purpose is to signify his own capabilities through his *work* and to confirm his solidarity with other workers or his family through *spending* his time this way.   (de Certeau, 1988: 25–6)

La perruque, therefore, is appropriation of existing resources for purposes for which they were not intended to be used. It creates new meaning for existing resources, thus opening possibilities for new social practice. The possibility of doing this may be tightly controlled, which leaves little room for improvisation, or there may be intentional slack in the system. A 'perfectly efficient' organization has full accountability and predictability and no slack, and therefore no space for new interpretations or practices. This is why strategic allocation of slack may sometimes be the most profitable investment in organizations that try to facilitate innovation (Tuomi, 1999: 366–7).

Organizational and social structures therefore often constrain the everyday practice of individuals. Individuals, however, also intelligently appropriate these constraints as resources, and manage their everyday life by building meaningful worlds out of the materials provided in the social system. According to de Certeau, individuals competently steer their life in the labyrinth of social constraints and trick existing powers to help overcome problems in everyday life.

The actual order of things is precisely what 'popular' tactics turn to their own ends, without any illusion that it will change any time soon. Though elsewhere it is exploited by a dominant power or simply denied by an ideological discourse, here order is *tricked* by an art. Into the institution to be served are thus insinuated styles of social exchange, technical invention, and moral resistance, that is, an economy of the '*gift*' (generosities for which one expects a return), an esthetics of '*tricks*' (artists' operations) and an ethics of *tenacity* (countless ways of refusing to accord the established order the status of a law, a meaning, or a fatality).   (de Certeau, 1988: 26)

De Certeau maintains that modern society is increasingly dominated by economy and economic powers. Yet, an alternative economy of gifts survives in the margins and interstices of this dominant system. This alternative economy, therefore, becomes an important space where new knowledge and practice can emerge. Indeed, as Linux and other open source projects have shown, new social spaces for creativity and collaboration can also form around information highways, not only in Rome or Naples. According to de Certeau, the constraints of modern society penetrate all social life but, partly because of this global influence of the constraints, the alternative economy also becomes increasingly global and prominent:

It is even developing, although held to be illegitimate, within modern market economy. Because of this, the politics of the 'gift' *also* becomes a diversionary tactic. In the same way, the loss that was voluntary in a gift economy is transformed into a transgression in a profit economy: it appears as an excess (a waste), a challenge (a rejection of profit), or a crime (an attack on property).   (de Certeau, 1988: 27)

According to de Certeau, scientific research is still characterized by this economy of gifts. Working with its machines and using scraps, scientists can divert organizational resources, create products that signify an art and solidarities, and play the game of free exchange. Work can, then, be creative and not just production for the ends defined by others.

In these ways we can subvert the law that, in the scientific factory, puts work at the service of the machine and, by a similar logic, progressively destroys the requirement of creation and the 'obligation to give.' I know of investigators experienced in this art of diversion, which is a return of the ethical, of pleasure and of invention within the scientific institution. Realizing no profit (profit is produced by work done for the factory), and often at a loss, they take something from the order of knowledge in order to inscribe 'artistic achievements' on it and to carve on it the graffiti of their debts of honor.   (de Certeau, 1988: 28)

At the level of individual everyday practice, innovation is therefore often materialized through improvisation and creative use of existing resources. The space of innovation exists in the periphery and on the margin, in under-specified, undetermined and unused spaces. Innovation can, however, also exist in interstices of current practices. At the level of social practice, innovation can occur when the relationships between social practices are reorganized. For example, tools that are produced for specific uses can be adopted for new uses. The linkages within an ecology of communities of practice can be recombined in new ways. Innovation can therefore have its source in the dynamic recombination of production and consumption relationships also at a level which cannot be reduced to the individual level of analysis. The space of recombination of social practice is qualitatively different from the space where *la perruque* and individual improvisation exist. Whereas individuals improvise within the constraints of social practice, communities can intentionally put practices into new contexts.

2.10 DYNAMICS OF NETWORKED INNOVATION SPACES

In the ecology of communities, innovation can often be detected as reorganization of relationships between different communities. For example, Bakelite producers can suddenly start to talk with car manufacturers (Bijker, 1997). Such reorganization can also be seen, for example, in the recent development of the travel industry, where traditional travel agents in many cases have been replaced by Internet services that enable travellers to directly reserve tickets, hotel rooms, and rental cars. This is a prototypical case of 'disintermediation' where some communities of practice simply become unlinked in the ecology of communities.

As was noted above, practices comprise artefacts and reproduced activity. Practices rely on an infrastructure of humans, communication, tools, materials, and products. A community of practice can therefore interface with other communities and its practices through various representations of practice. Sometimes practice is represented by a person who stands for the community. Communities may, however, also be represented by artefacts that they produce. In other words, a product can become a resource that stands for the capabilities of the community that creates it. In the evolution of a network of practices, services therefore can become products. Indeed, there is no fundamental distinction between them.

A simplified schematic example of evolution of a network of innovation is shown in Fig. 2.1. The figure shows some main linkages around a computer operating system development community as they existed in the 1970s and at the end of the 1990s.

At the end of the 1970s, computer operating system programmers lived in a world where their professional practice was relatively well defined. The programmer wrote code using a text editor and used either a compiler or an assembler to generate binary code for the target computer. At the end of the 1990s, the environment had become more complex. The 'computer' itself had become transformed. It had become an increasingly complex network of components, each with their own development communities. One of the underlying technologies, the Internet, had become a major use of the computer, and programmers tried to manage the increasing complexity of new and continuously changing components using Internet newsgroups.

As sociologists have often argued, the development of modern society creates increasingly complex interrelationships (e.g. Simmel, 1990; Halbwachs, 1992; Berman, 1982; Giddens, 1990). Part of this complexity, however, is contained by creating interfaces that hide some of this complexity. Actor-network theorists, for example, argued that complexity needs to be 'black-boxed' and 'translated' for practical action to proceed (Callon, 1987; Latour, 1999; Law and Hassard, 1999). We can talk about the 'postal service' or 'the British government' without knowing in detail what processes, people, and technologies these black boxes actually contain (Law, 1992). For most purposes, the postal service can be represented by a person behind the counter and some simple artefacts, such as an envelope and a stamp. Similarly, we can talk about 'a computer' without knowing exactly what its components are or how they have been connected. More generally, society, and social meaning itself, can be viewed as an order which simplifies this complexity (Luhmann, 1995; 1990).

At each point of time, a society consists of a complex interlinked set of social practices, some of which are represented as technological resources. Practices evolve continuously, and generate tensions in the process. As was noted above, this

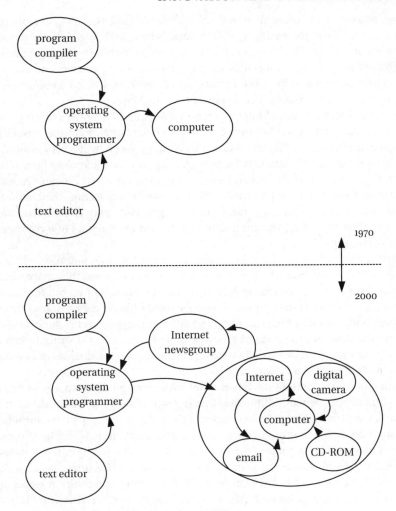

Fig. 2.1. Changing production–consumption linkages

is an important driver for innovation. Technologies and their improvements may emerge as expressions and articulations of ways that these tensions can be released and reduced. However, in addition to such continuous evolution and articulation, social practices—and resources they generate—can also be recombined. When the ecology of communities becomes sufficiently complex, opportunities for such recombinatorial innovation increase. This process is similar to *la perruque*, although it occurs at the level of an ecology of social practices.[15] Recombinatorial innovation, therefore, relies on appropriation of *unintended resources*.

---

[15] The term recombinant growth has been used in economics to describe growth that is produced by combining technologies in novel ways (cf. Varian, 2000). These models originate from theoretical

The dynamics of recombinatorial innovation depend on many factors. One important factor is the mobility of resources. When there is only little mobility, a community is fixed in its network of relationships. For example, a producer of technology may produce tools only for a given community of users and consume raw materials produced by a well-defined set of suppliers. In such a case the community may find itself stuck with its current relationships.

Mobility, therefore, defines how easy it is to reorganize the motives and meanings in a given community so that they can be linked with new practices. Mobility of resources often reflects characteristics of technology. When technology is architecturally rigid and inflexible, it may be difficult to appropriate it for novel purposes.[16] Mobility can, however, also depend on the system of values and meanings within the community. For example, members of a community define their identities in relation to the community (Lave and Wenger, 1991) and those socialization processes that reproduce the community often limit the mobility of a community (Engeström, 1987).

The community can also be uninterested in change. In some communities novelty is not valued. Sometimes the whole ecology may share values that make change difficult. Mokyr (1990), for example, argued that one important reason for the very different technological trajectories of European and Chinese cultures was that after the thirteenth century China was in many ways a homogeneous Confucian system. The power structure in China made it possible to get rid of disturbing change, and—in contrast to Europe—there were no independent small duchies or city-states where deviants and people with new ideas could flee (Mokyr, 1990: 236).

Even when the community would want to reorganize its relations within the web of interdependences, it may find that the web is sticky. Recombination at the network level is about relational change. It does not happen only in one focal community. Indeed, one may argue that usually communities have a very limited capability to make non-local innovations, and that innovation typically emerges either at the level of individual creativity within a community of practice, or as a result of inter-community tensions that force reorganization of inter-community relationships. The following chapters, however, show that communities can create interfaces that facilitate recombinatorial innovation also at the network level.

biology and models of evolution. In contrast to most current discussions on recombinatorial growth, I use the term in the context where 'the components of a technology' cannot be taken apart in any objective sense. Recombination is always linked to change in social practices. The unit of analysis is therefore not a decomposed artefact, but a complex system of humans and artefacts. In this sense, the usage is actually quite close to the original biological analogies that inspired the theory of recombinatorial growth. Eigen and Schuster (1979), for example, proposed a model of hypercycles to describe evolution as increasing complexity of linkages between subsystems. I have discussed the hypercycle model in detail in Tuomi (1982).

[16] This problem is implicitly addressed by product development models that try to build flexibility into the development process. As was noted before, on a more fundamental level the question, however, is not only about keeping development options open as long as possible but about producing technical designs and architectures that explicitly assume interpretative flexibility and unintended uses.

Recombination is difficult because it is relational, but advances in information and communication technologies also make it more possible than it was before. This is one reason why the 'downstream' is becoming increasingly important in innovative change. Downstream innovation implies that an existing technological opportunity is taken into a novel use. As this process requires change only in the practices of the downstream community, innovation can be local and there is no need to synchronize change across communities.

It is important to note that radical innovation can occur gradually, without revolution. Internet email is an example here. It emerged as a result of a low-profile skunk-work project, was taken to many illegitimate uses by the developer communities of the Internet, and eventually redefined the Internet itself. Throughout this process, the basic technology of email has remained quite stable. Only relatively late, actors such as Microsoft or AOL have emerged as important players in the development of email. Eventually, of course, the reorganized relationships may have considerable impact on the identity of the members of a specific email-using community and its practices, as well as in the direction of evolution of the community.[17]

Another important factor in the dynamics of combinatorial innovation is transparency. Community resources are usually represented to the 'outside world' through translation mechanisms such as products and persons who stand as proxies for the community. These translation mechanisms link existing communities, and therefore they reveal the inner structure of the black boxes in a way that is appropriate for the given interrelationship. To create new linkages, black boxes often have to be opened and new ways to create resources have to be defined. If there is

---

[17] The process is therefore essentially the same as underlies biological evolution of functional organs. Many scientists, since Darwin, have been puzzled by the fact that complicated systems such as the human eye can emerge in the process of evolution. The answer to this puzzle is that the eye does not emerge *as* an eye. After photosensitive cells emerge they are appropriated for new uses by the animal in question, thus creating a new developmental path where 'vision' becomes relevant. When vision has become relevant, the diverse possibilities of seeing can develop in parallel and very different biological architectures can develop for seeing. The traditional Darwinistic models of evolution miss this point, and consequently evolutionary descriptions of technological change tend to assume that variation and selection processes are the key to technological development (e.g. Ziman, 2000). The 'blind' Darwinistic variation tries to get rid of determinism through randomness, and sees the generation of 'useless' and potentially 'failing' variations at the core of evolution. A more interesting possibility is to view evolution as a 'creative' process where new domains of being emerge with the capability to operate in new ways. This, of course, was exactly what Bergson (1983) argued at the beginning of the last century. Variation, in this view, is not generated by more or less useful versions of a given characteristic; instead, existing capabilities are appropriated for qualitatively new uses. Although selection plays a role in evolution, the Darwinian model misses this creative domain of evolution. In technology development this creative domain is essential.

Bergson also noted that it is difficult to describe how functional organs emerge as evolution is a continuous process. It is difficult to explain the emergence of the eye, for example, because at the point of time when something becomes an eye there doesn't exist any function of seeing (cf. Maturana and Varela, 1988). All our explanations of the eye, therefore, tend to be retrospective projections into a world where vision didn't exist as we now know it. A similar retrospective projection often underlies theories of innovation, as will be shown in a later chapter.

no way to penetrate the complexity that is hidden behind the existing translation mechanisms it is difficult to invent new ways to combine resources. Transparency therefore defines how easy it is for outsiders to see whether such new linkages could make sense for the community. It also helps the outsiders to create their own translation processes in ways that are easy to link with other communities.

A third important factor in mobility is the capability to create mechanisms that hide complexity and translate communities into resources. Creation of a technological product is itself an important way of doing this. For example, computer users do not need to integrate their activities with the complex processes that design, develop, and manufacture computers and their components; instead, they can simply buy a box that embeds the various competences that are needed to make computer programs operational. Effective ways to hide complexity facilitate mobility and recombination of resources. Market economy, in general, is such a mechanism. Other similar generic mechanisms include standardization of technical interfaces, production of standard components, and, for example, standard formats for information exchange.

As the following chapters show, one common way to create transparency is through individuals who are simultaneously members of several communities. Information technology, however, has also created new ways to increase transparency. The open source development model, for example, explicitly utilizes these new opportunities to create new technology. Indeed, the open source model relies on mechanisms that effectively support recombinatorial innovation.

In modern society mobility of resources has become a central value. Producers and consumers form networks that are in constant transition (Berman, 1982; Harvey, 1990; Castells, 1996). Information technology, in turn, has increased mobility of resources and competences. Moreover, information technology is exceptional because it can simultaneously both create transparency and hide complexity, and dynamically change the visibility of relationships in an ecology of communities. At the level of ecology of communities of practice, the speed of recombination has therefore increased. The space of recombination has become an increasingly dominant space for innovation.

The viability of this space, of course, depends on someone creating new knowledge and competences in the first place. For example, today Silicon Valley has become a portal to the global innovation system, specializing in the rapid recombination of resources and technologies. The viability of this space, as a social space, is an open question. As the following chapters show, many of the innovations that underlie entrepreneurial and innovative activities in Silicon Valley were created a long time ago, often far from Silicon Valley. In this sense, Silicon Valley may be a niche player in the innovation game. Innovation does not depend only on the creativity of individuals, their capability to solve existing problems, or recombination. It also depends on time and place for improvisation and play, as well as the existence of slack.

In the following chapters, these concepts are made more concrete by studying how important technological innovations have been generated. I will introduce several historical examples of new technologies and use them as empirical cases to study innovation processes, elaborating and clarifying the above discussed concepts in the process. The focus in the following chapters will be on innovations that relate to the Internet. Although this is a relatively narrow focus, it also helps us to understand some innovation processes that are central in the 'new economy'. Furthermore, this focus also highlights the ways innovation processes are changing as information and communication technologies become widely used to support and facilitate innovation. I will start with a brief description of one of the most influential and visible innovations of the last decades: the invention of the World Wide Web.

Today, it seems clear that the World Wide Web became one of the most important technological developments around the end of the millennium. It is therefore useful to revisit the history of the World Wide Web and try to understand who were the actors that made the World Wide Web happen; and why, indeed, it became such an important part of our life.

3

Inventing the Web

According to most accounts, the World Wide Web was created when Tim Berners-Lee programmed its first version at CERN, November 1990. For example, in the foreword for Berners-Lee's own history of the World Wide Web, Michael Dertouzos writes:

Amid the barrage of information about the World Wide Web, one story stands out—that of the creation and ongoing evolution of this incredible new thing that is surging to encompass the world and become an important and permanent part of our history. This story is unique because it is written by Tim Berners-Lee, who created the Web and is now steering it along exciting future directions. No one else can claim that. And no one else can write this—the true story of the Web.   (Dertouzos, in Berners-Lee and Fischetti, 1999: p. ix)

As Dertouzos notes, the World Wide Web has become an important part of our life. But how, exactly, did this happen? When was the Web invented? Do we really find a heroic innovator, as Dertouzos implies, when we study the records of the information age? How are such major innovations produced?

According to Berners-Lee, he first had the idea of creating a computer system that could link pieces of information when he came home from high school and found his father working on a speech. His father was member of a team that had programmed the world's first commercial stored-program computer, the Manchester University 'Mark I', in the early 1950s. While preparing the speech, his father was reading books on the brain, looking for clues about how to make a computer intuitive and able to complete connections as the brain did. Berners-Lee discussed the idea with his father, and 'the idea stayed with me that computers could become much more powerful if they could be programmed to link otherwise unconnected information' (Berners-Lee and Fischetti, 1999: 4).

According to Berners-Lee, the idea stayed in his mind during his studies at Queen's College at Oxford University and remained in the background until 1980, when he took a brief software consulting job at CERN. During this visit to CERN, Berners-Lee wrote a program to help him remember the connections among the various people, computers, and projects at the lab. He called the program Enquire, short for Enquire Within Upon Everything, 'a musty old book of Victorian advice' published in 1856. Berners-Lee's parents had the book on their bookshelf. It served information on practical matters such as how to remove clothing stains and tips on investing money.

Enquire was developed to solve a practical problem. CERN is a large international research institution, with thousands of people in its phone book. Most of them are scientists who come to CERN for short periods of time to do their research and develop systems for experiments. According to Berners-Lee:

The big challenge for contract programmers was to try to understand the systems, both human and computer, that ran this fantastic playground. Much of the crucial information existed only in people's heads. We learned the most in conversations at coffee at tables strategically placed at the intersection of two corridors. I would be introduced to people plucked out of the flow of unknown faces, and I would have to remember who they were and which piece of equipment or software they had designed.   (Berners-Lee and Fischetti, 1999: 9)

In his spare time, Berners-Lee tinkered with Enquire. Once the system was working, he started to use it to keep track of who had written which program, which program ran on which machine, and who was part of which project. In Enquire, information was stored as a 'node' in a linked network. Each page of information was a node. A new node was created by creating a link from an existing node. The links would show up as a numbered list at the bottom of each page, and the only way of finding information was by browsing the links, starting from the start page. The links between pages that were stored within a single file worked both ways, so that it was possible to navigate from any node to its associated nodes. It was, however, also possible to navigate across different files. Such 'external links' were one-way links. According to Berners-Lee (1999: 10), this design tried to avoid problems with maintaining links when many people used the system and created their own pages. If all links would have been two-way links, anyone could have added associations to an existing page, thus potentially creating a large set of changing links that needed to be maintained by the system.

Berners-Lee left CERN after six months and went to work for a friend who had started a business. He wrote programs that made it possible to use dot-matrix printers to print graphics and use the printer as a typesetting machine. In the process, Berners-Lee and his colleagues wrote a 'markup language' which controlled the way a document was printed.

In 1983 Berners-Lee decided to apply for a fellowship at CERN. He moved to CERN in September 1984, to work with the 'data acquisition and control' group, which was responsible for capturing and processing the results of experiments at CERN. As a departure gift from his previous job, he got a portable Compaq personal computer.

While at CERN Berners-Lee rewrote Enquire so that it would work both on his portable Compaq and the minicomputer he was using at CERN. According to Berners-Lee, it was clear to him that CERN would need something like Enquire:

In addition to keeping track of relationships between all the people, experiments, and machines, I wanted to access different kinds of information, such as a researcher's technical papers, the manuals for different software modules, minutes of meetings, hastily scribbled notes, and so on. Furthermore, I found myself answering the same questions asked frequently of me by different people. It would be so much easier if everyone could just read my database. (Berners-Lee and Fischetti, 1999: 15)

Although Enquire provided a way to link documents, the system had the problem that different computers used different operating systems. Berners-Lee, however, had just programmed a general 'remote procedure call' (RPC) program for CERN. RPC programs enable programs running on one machine to call programs on another machine even when the computers use different operating systems. Berners-Lee hoped that he could somehow combine Enquire's external links with interconnection schemes he had developed for RPC (Berners-Lee and Fischetti, 1999: 16).

The basic concept in Enquire was to build a network of linked documents. The user could move from one document to another linked document simply by 'following the link'. At the time Berners-Lee was developing his Enquire, this concept was well known. Indeed, the idea had been widely discussed since 1945, when Vannevar Bush had published an influential article that described a machine that could be used to store and retrieve documents using such links (Bush, 1945). In the 1960s, it became known as 'hypertext'. The term was coined by Ted Nelson in his presentation at the Association of Computer Machinery Conference in 1965. During the next decades Nelson actively promoted the idea that hypertext could become a new interactive medium where documents and digital content could be stored in a distributed global 'docuverse'.

The first working hypertext system was developed at Brown University in 1967 by a team led by Adries van Dam (Naughton, 2000: 220; Conklin, 1987). A major breakthrough in hypertext occurred in August 1987 when Apple launched its HyperCard program. HyperCard had graphical 'cards' where the user could put 'hot spots'. These hot spots linked cards to each other. When the user clicked a hot spot with a mouse, the system displayed the linked card. A collection of linked cards was called a 'stack'. The stack had a 'home' card that provided the base node from which the user could start traversing the stack. Although at first HyperCard was designed to enable the user simply to jump from one card to another using hot spots, early on hot spots were made more general. A language called HyperTalk was designed to enable the user to define what happened when a hot spot was clicked (Naughton, 2000: 222–8).

3.1 THE FIRST WorldWideWeb PROPOSAL

In 1988 Berners-Lee was reading about hypertext, and learned about Ted Nelson's ideas of generating a global network of documents (Berners-Lee and Fischetti, 1999: 65). At the end of the year he discussed the idea of a CERN hypertext system with his boss Mike Sendall. Sendall suggested that Berners-Lee should write a project proposal. During the next months Berners-Lee wrote a paper that described his idea and tried to persuade CERN management to develop a distributed hypertext system. The paper was entitled 'Information Management: A Proposal' (Berners-Lee, 1990).

The proposal focused on the problem of losing information at CERN. CERN is a large project organization, where research teams come together for various periods of time. The research done at CERN generates very large amounts of data and documents. According to the proposal, the high turnover of people creates a problem:

When two years is a typical length of stay, information is constantly being lost. The introduction of the new people demands a fair amount of their time and that of others before they have any idea of what goes on. The technical details of past projects are sometimes lost forever, or only recovered after a detective investigation in an emergency. Often, the information has been recorded, it just cannot be found.   (Berners-Lee, 1990)

Although CERN was in many ways a special environment, Berners-Lee also predicted that 'CERN is a model in miniature of the rest of world in a few years time'. He noted that in ten years' time there would be many commercial solutions for the information management problems but argued that there was also a need to do something to help to alleviate the current problems.

The basic idea in Berners-Lee's proposal was to build a system where new documents could be flexibly added. To enable different types of information to be stored and retrieved, it was important that the method of storage did not restrict the types of information that were to be stored. Whereas hierarchical document management systems and keyword-based retrieval assumed a pre-defined categorization of documents, Berners-Lee proposed that the system should simply consist of linked documents. A practical requirement at CERN was that the system should enable remote access across networks with computers that used different operating systems. The system was also to be non-centralized so that new documents could be added without central control or coordination. In addition, the system was to support retrieval of data in existing systems so that it would be useful early on and that a critical mass of users could be recruited to put useful content on the system. The users should also be able to create their own private links and annotations to and from public documents. It should also be possible to create 'live' links that would retrieve data dynamically when a hot spot was clicked.

In his proposal, Berners-Lee also noted that such a network would provide interesting opportunities for automatic analysis. It could be possible, for example, to detect software stored in the system which had no documentation, or organizational divisions that had no people. By analysing the topology of the network one might also be able to make conclusions concerning the ways an organization or a team should be managed.

The system was supposed to be a simple one, without much in the way of bells and whistles. According to Berners-Lee, storage of ASCII text and displays with 24 lines of 80 character lines would be sufficient in the short term. In 1989 many users at CERN had character-based terminals, and Berners-Lee argued that graphical user interfaces were not important for the time being. Although Berners-Lee noted that earlier hypertext systems had tackled the problems of copyright and data security, he maintained that at CERN these were of secondary importance. To the extent that access control was needed, the users could simply rely on access control provided by the file system of the underlying operating system.

In his proposal Berners-Lee noted that many document management systems used a centralized database, and that there were few systems that took Ted Nelson's idea of a distributed document network seriously. Partly this was because work in hypertext and hypermedia was focused on publishing hypermedia information, for example, using optical discs. According to Berners-Lee, Digital's 'Compound Document Architecture' was an interesting attempt to develop a distributed document architecture, and according to rumours Digital was going to extend it for hypermedia use. An important incentive to develop hypermedia systems was provided by the US Department of Defense, which had started to require that its contractors provided documentation in the SGML markup format.

Already from this brief description, it is easy to see some key characteristics of the process that led to the World Wide Web. In his proposals, Berners-Lee emphasizes the point that it is impossible to predict how the users will use the system. Instead of proposing a system that would address all the difficult requirements of a full document management system—for example, information security and document version control—the plan consists of a simple platform that facilitates sharing of information across existing systems. In the process he also relies on existing ideas and previous work, for example, the SGML standard for text markup and the Internet remote procedure mechanisms. Although Berners-Lee believes that the system would become a tool for document management, the design actually leaves out almost all document management functionality. Much of this functionality is expected to come from the proper use of the system. Instead of coding sophisticated security mechanisms in software, for example, Berners-Lee assumes that common sense tells the users what kind of documents they can put into the system—and even when this is not true this is not a major problem as the institutional role of CERN in any case requires open communication.

The World Wide Web is a radical innovation when viewed through the traditional document management perspective. In many ways its underlying assumptions do not make sense. For an information systems manager, the system is a disturbing thing. For example, it cannot be managed in conventional ways. Although the system naturally enters the domain of responsibility of information system management, it has no well-defined points of control. From an information system designer's point of view, on the other hand, the system is also radical. It off-loads much of the functionality to the users. It tilts the balance between machines and humans, and shows that information systems cannot be understood simply as technical systems. At the same time, it shifts the balance of power. The World Wide Web is a radical innovation as it requires revolution in organizations and their document management traditions. Perhaps because of this, the World Wide Web takes off first outside organizations and as unofficial skunk-work projects within organizations. Here a new technology opens new domains of freedom. As a result, innovative users rush to construct their own worlds and make their own marks in this novel space of anarchy.

3.2 STATE OF THE ART: KMS

Traditionally, innovation is supposed to be something novel and it should improve on existing technology. Was the World Wide Web, in this sense, an innovation?

Berners-Lee's first proposal lists a number of references, mentions that there exists a newsgroup that discusses topics on hypertext, and notes that there have been conferences on hypertext where the state of the art has been represented. The proposal explicitly refers to a book chapter written by Ted Nelson in 1967, and to four articles in a special issue of the Communications of the ACM, published in July 1988. This special issue contained papers from the Hypertext '87 conference, held at University of North Carolina in November 1987. This was the first major conference devoted to hypertext (Smith and Weiss, 1988).

In their introduction to the Communications of ACM special issue, the guest editors define hypertext as a form of electronic document. More precisely:

... hypertext is an approach to information management in which data is stored in a network of nodes connected by links. Nodes can contain text, graphics, audio, video, as well as source code or other forms of data. The nodes, and in some systems the network itself, are meant to be viewed through an interactive browser and manipulated through a structure editor. While the term, hypertext, was coined by Ted Nelson during the 1960's, the concept can be traced back to Vannevar Bush's description of 'the memex'.   (Smith and Weiss, 1988: 816)

Several different systems that implemented the hypertext concept were described in the conference and reported in the special issue. These included InterMedia, developed at Brown University, where Adries van Dam had built the first hypertext editing system in 1967; NoteCards, an ambitious and sophisticated system that had been developed at Xerox PARC; Symbolics Document Examiner, which provided graphical online access to hypertext user documentation for the users of Symbolics Lisp-machines; Neptune, a hypertext system for computer-assisted software engineering; and WE, a hypertext authoring system developed at the University of North Carolina that produced both paper and electronic documents and which modelled human cognitive processes.

One of the systems described in the special issue, and referenced by Berners-Lee in his CERN proposal, was KMS, short for 'Knowledge Management System'. It was a distributed hypermedia system that was based on an earlier system, ZOG, developed at Carnegie-Mellon University, starting in 1972 (Akscyn, McCracken, and Yoder, 1988). KMS stored its documents in a distributed database, and represented documents as frames:

The heart of KMS is its conceptual data model. A KMS database consists of screen-sized WYSIWYG workspaces called frames which contain text, graphics and image items. Individual items can be linked to other frames or used to invoke programs. The database can be distributed across an indefinite number of file servers and be as large as available disk space permits. The KMS user interface employs a form of direct manipulation designed to exploit a three-button mouse. A combined browser/editor is used to traverse the database and manipulate its contents. Over 90 percent of the user's command interaction is direct—a single point-and-click designates both object and operation. Running on Sun and Apollo work-stations, KMS accesses and displays frames in less than half a second, on average.   (Akscyn, McCracken, and Yoder, 1988: 820)

The central metaphor in KMS was that the database is a 'universe of connected spaces through which users rapidly travel, like pilots navigating spacecraft in the real universe' (Akscyn, McCracken, and Yoder, 1988: 822). By pointing and clicking an item, the user could move to the frame linked to the item. The user could also directly manipulate the content of the frame at any time. New linked frames could be created by pointing to an unlinked item and clicking the left mouse button. Clicking an item could also invoke a program, for example, to automatically print a paper-version of a specific part of the network.

The article by Akscyn, McCracken and Yoder discusses several design challenges for distributed hypertext systems. One question was what kind of nodes the system should support. Some systems, such as NoteCards and HyperCard managed the links as integral parts of the node document, whereas other systems, such Inter-Media, had a separate database for links. KMS used the first approach. It also relied on a spatial metaphor, using frames as 'spaces' where content and links could be placed. A node, therefore, could exist even when it had no content. In contrast to some other systems, such as NoteCards that had different node types, KMS had only a single type of node, the frame. Different types of content—for example images and text—could be embedded within a frame but all nodes were of the same type.

KMS used individual text items as links, and displayed a small circle before them to notify the user that the item was a link. Links were separated from the main content in a frame. Whereas other hypertext systems usually embedded links directly into the main content, as the World Wide Web does, KMS authors argued that by separating links and content it was, for example, easier to use descriptive names for the links. This made it also easier to arrange links so that a coherent paper document could be printed by automatically processing the linked frames.

Whereas some hypertext systems supported 'typed' links—for example, to note that a linked document was a 'counterargument'—KMS supported only two types of links. Links could be either *tree links*, which connected higher level frames to lower level frames in a hierarchy, for example chapters of a book to sections, and sections to paragraphs; or links could be *annotation links* that simply associated two frames. In contrast to NoteCards and HyperCard, KMS links were one-way links. This was an important design choice. Instead of storing in a frame information about frames that linked to it, KMS simply supported a backtrack command. With the backtrack command the user could always return to the previous frame, even when two-way links didn't exist. The research by the KMS authors showed that backtracking was often used several hundred times per hour. KMS could retrieve the previous frame on average in 0.7 seconds. The authors also considered several other design issues, including the necessity of providing graphical views of the underlying document network, ways to avoid 'getting lost in hyperspace', search mechanisms, joint authoring, access control, and support for communication and group annotation.

One technically and architecturally challenging problem in joint authoring systems is how several people can access and edit shared documents without simultaneously making incompatible changes in the same document. Indeed, this is one of the key reasons why document management systems are difficult to implement without some form of centralized control. Usually this problem has been solved by 'locking' those documents which someone is currently editing. When other users try to edit a document that is currently being edited by someone else in a typical document or database management system, they cannot do it. Management of such locks, however, means that the system has to maintain information about the status of each document. The problem of locking documents, therefore, normally leads to a centrally coordinated system architecture. As the content in KMS was divided into relatively short frames, and the assumption was that there were much more frames than users, the KMS architecture was based on the idea that the probability of two persons trying to edit the same frame simultaneously was low. Therefore the system did not have to 'lock' frames to prevent interference between users. In those rare cases where collisions occurred, the users could handle them manually. The KMS authors also noted that if the risk of collisions was high, the users could use informal coordination to reduce the problem, for example, by placing text on a frame that warned other users that the frame was being edited (Akscyn, McCracken, and Yoder, 1988: 831–2).

3.3 ARCHITECTURE OF THE WORLDWIDEWEB

Implementation of a generic vision, such as a distributed platform for linking documents, requires solving a set of specific design problems. Visions, however, are rarely implemented in their original form in real life. During the implementation process, visions change. The priority and importance of technical problems therefore depends also on the actual order of project implementation. When the order of implementations tasks is changed, this can have drastic implications for the success of the project at hand. Depending on the path taken, the end result also may look quite different.

Although Berners-Lee and the KMS development group, for example, shared very similar overall goals, KMS had a much stronger focus on collaborative use of the system and the problems of content creation. Berners-Lee, in contrast, saw the primary problem as a problem of accessing existing data. As data existed in many different formats and on many different machines in CERN, the primary problem was about translating this data into formats which different terminals and workstations could display, and to provide unified access to stored data. As all the different systems managed their data and access rights in different ways, the only realistic assumption was that Berners-Lee's system would only deal with information that was freely available. Therefore the problems of security and privacy were not issues for Berners-Lee.

Similarly, the problems related to authoring and joint editing were not issues at the first phase of Berners-Lee's proposed project. Berners-Lee had a pressing need to prove that the system he proposed would be useful, and the easiest way to do this was to provide access to existing content. The system facilitated combination of existing information resources by translating them so that the users could do something with these resources. This, of course, depended to a large extent on the fact that there already existed such information sources.

When we compare Berners-Lee's proposal and KMS, the main difference in the architectures is that KMS was based on a shared database, whereas Berners-Lee's proposal relied on an architecture where a heterogeneous collection of databases and files could be used as document storages. Although the KMS database was not centrally managed, didn't require a central locking system to prevent conflicting edits, was physically distributed across many computers, and could be accessed anywhere from the network, the system still used a master file that contained information on the location of frames in the system. KMS therefore, for example, allowed its users to manage versions of documents by 'freezing' and storing a specific state of the system. KMS also explicitly supported access control for each frame, so that the 'owner' of a frame could decide whether the frame was editable or visible to other users. Although KMS avoided the technical challenge of locking documents to prevent collisions in editing, it effectively reintroduced locking to manage access control. The locking mechanism, however, relied on the access control mechanism of the underlying distributed file management system. This required that all documents were stored within a single file system.

In his original proposal, Berners-Lee doesn't tell how he would solve problems related to locking or version management. Indeed, he did not mention these problems at all. He did, however, assume that access control would probably not be a problem at CERN. Instead of building access control to the proposed system, the users could simply rely on the existing access control of the file system they were using. A new proposal, written in November 1990 by Berners-Lee and his colleague Robert Cailliau, explicitly noted that the project would not aim to use any sophisticated authorization systems. As long as a document was shared only within the users of a specific file system—for example, users of a specific computer network at CERN—the file system access control mechanism could be used to limit access to the document:

Data will be either readable by the world (literally), or will be readable only on one file system, in which case the file system's protection system will be used for privacy. All network traffic will be public.   (Berners-Lee and Cailliau, 1990)

Berners-Lee's concept was based on the assumption that the architecture should clearly separate browsers, which supported navigation, and servers that stored content. One central idea was that there could be many different kinds of servers. For the KMS developers this probably would have been an unnatural idea. KMS was a commercial system, developed and sold by Knowledge Systems, Inc., a US firm that was set up in 1981 to commercialize the ZOG hypertext system developed at Carnegie-Mellon University. ZOG and KMS were used for applications such as computer-assisted management of aircraft carriers and nuclear power plants. In Berners-Lee's vision, different servers could be combined without anyone being able to control the evolution of the system. The lack of control meant that it would have been difficult to sell Berners-Lee's system, or provide well-defined services for it.

In essence, the WorldWideWeb was to be a translator that served different types of content, such as program documentation, help files, telephone directories, and newsgroup postings, to the user in a unified from. The way information was to be displayed was negotiated by the server program and the browser program when the user accessed the information. Whereas KMS was based on the assumption that the system was to support the creation and management of hypertext documents, the WorldWideWeb was largely based on the assumption that it would be useful to provide unified access to many different information sources. Although Berners-Lee clearly intended the system to eventually become a system where users could create their own documents and where reading and editing would be supported by the same browser (Berners-Lee and Fischetti, 1999: 70–1), the project proposals promised to add editing capabilities later, after access to existing documents had been solved.

Although Berners-Lee's original vision was to create a system where users could develop their own content, the existence of content at CERN made it possible to organize the implementation plan so that the first phase would simply start by providing access to existing content. This meant that the first version of the system could be implemented by building a browser and some gateway programs that translated existing data into a format that the browser could display. Berners-Lee, therefore, was able to avoid almost all difficult technical problems that earlier systems and their developers were trying to solve.

Indeed, many of these technical problems are still unsolved in the World Wide Web. Instead of solving them, the users have found ways around them. The simple architecture of the World Wide Web has made it possible to develop many complementary systems that alleviate the most acute problems. For example, search engines and security fire walls are being widely used today.

3.4 MOBILIZING RESOURCES

If experts had evaluated Berners-Lee's proposals at the end of 1990, it is quite probable that they would have judged them as technically inadequate and lacking novelty. For example, the WorldWideWeb development project most probably would not have got funding from European research initiatives or from the US National Science Foundation. Existing commercial systems had long provided more advanced hypertext functionality and the WorldWideWeb simply wasn't a very good document management system. For CERN, however, Berners-Lee's proposal was not about creating novel technology. The proposed system was a solution to a perceived problem of accessing information in different systems. Technical sophistication or novel ideas were not interesting as such. Indeed Berners-Lee emphasized that the project was to be a very pragmatic implementation project where existing products, such as KMS, were to be used when possible.[1] But when Berners-Lee distributed his proposal at CERN, at the end of March 1989, there was no response:

I gave it to people at a central committee that oversaw the coordination of computers at CERN. But there was no forum from which I could command a response. Nothing happened. (Berners-Lee and Fischetti, 1999: 22)

During 1990 Berners-Lee read more about hypertext and got increasingly interested in the Internet. He had first learned about Internet technology when he programmed the RPC, and the Internet TCP/IP protocols seemed to provide a way to connect the different machines used at CERN into a unified network. According to Berners-Lee, at that time the Internet was relatively little known in CERN. It was, however, rapidly becoming available for many operating systems used there.

As Berners-Lee didn't receive any response to his original proposal, he reformatted it and distributed it again at CERN, in May 1990.[2] Again the proposal was shelved (Berners-Lee and Fischetti, 1999: 22). At the same time, however, Berners-Lee was discussing with his boss about the possibility of buying a NeXT personal computer, which had a lot of intriguing features and which supported effective development of graphical user interfaces and object-oriented programs. Berners-Lee was able to convince his boss that this new machine would be a good platform for learning about object-oriented systems. Berners-Lee's hypertext program therefore became an experiment in using the NeXT operating system and its NeXTStep programming tools.

The NeXT computer was the most sophisticated personal computer available in 1990. It was created by a team of designers led by Steve Jobs, the founder of Apple Computers, who had launched NeXT Inc. after leaving Apple. The architecture of NeXT was designed to support rapid object-oriented software development and it had strong support for developing graphical user interfaces and hypertext programs. Using the NeXT as a development platform, Berners-Lee started to write the code for his system in October 1990.

My first objective was to write the Web *client*—the program that would allow the creation, browsing, and editing of hypertext pages. It would look basically like a word processor, and the tools on the NeXT system, called NeXTStep, were ideal for the task. I could create an application, menus, and windows easily, just dragging and dropping them into place with a mouse. The meat of it was creating the actual hypertext window. Here I had some coding to do, but I had a starting place, and soon had a fully functional word processor complete with multiple fonts, paragraph and character formatting, even a spellchecker! No delay of gratification here. Already I could see what the system would look like.   (Berners-Lee and Fischetti, 1999: 28)

---

[1] The 1990 project proposal written by Berners-Lee and Cailliau required funding to purchase licences for KMS and Owl's Guide hypertext system, which was the first widely used hypertext development environment, introduced in 1986.

[2] According the Berners-Lee, he didn't make any changes to the content of his original proposal, which he first distributed in March 1989. The available May 1990 version, however, discussed a conference held in January 1990. The proposal also mentions that the project would provide an excellent opportunity to experiment with object-oriented programming techniques. It seems, therefore, that at least some modifications were done in the content of the proposal after it was first distributed.

To create hypertext links, Berners-Lee had to find a way to differentiate links from normal text. He studied the implementation of the NeXTStep text editor and found out that its developers had left some free bits for experimentation. Berners-Lee used this free memory space to store a pointer to the address of the node that was linked to the specific piece of text:

With this, hypertext was easy. I was then able to rapidly write the code for the Hypertext Transfer Protocol (HTTP), the language computers would use to communicate over the Internet, and the Universal Resource Identifier (URI), the scheme for document addresses.   (Berners-Lee and Fischetti, 1999: 28–9)

Berners-Lee had his first browser ready mid-November, and he called it WorldWide-Web. By December the browser was working with the Hypertext Markup Language (HTML). Berners-Lee also wrote the first server program for his NeXT, and programmed the browser so that it could also access files using the Internet file transfer protocol (FTP) and display articles in Internet newsgroups. As Berners-Lee recalls: 'In one fell swoop, a huge amount of the information that was already on the Internet was available on the Web' (Berners-Lee and Fischetti, 1999: 30). When his colleague Robert Cailliau bought another NeXT machine, Berners-Lee and Cailliau were able to access their HTML documents across the Internet. The system was working by Christmas Day 1990.

   Although Berners-Lee and Cailliau now had a working prototype of the WorldWideWeb, which they could demonstrate, they still had difficulties in convincing people at CERN that the system could be useful. To keep the informal project going, Berners-Lee had to find some way that users would see the potential of the system. One problem was that the system used the NeXT computer, which limited the spread of the system as other CERN computers were incompatible with NeXT. Berners-Lee therefore considered the possibility of reprogramming WorldWideWeb on a standard PC. This, however, would have required a lot of work and there was no guarantee that it could ever have been successfully done, given the limited resources available. Another way to increase the usefulness of the WorldWideWeb would have been to promote its use outside CERN. This was a problem as CERN was paying the salaries for Berners-Lee and Cailliau and it was not obvious why CERN should fund systems for the rest of the world. A third possibility was to use the WorldWideWeb to solve some actual problems at CERN. Although Berners-Lee himself saw the WorldWideWeb as a solution to a generic problem, to justify his work at CERN he had to find legitimate reasons for developing the system and using resources:

My head reminded me, however, that to attract resources I also needed a good, visible reason to be doing this at CERN. I was not employed by CERN to create the Web. At any moment some higher-up could have questioned how I was spending my time, and while it was unusual to stop people at CERN from following their own ideas, my informal project could have been ended. However, it was too soon to try to sell the Web as the ultimate documentation system that would allow all of CERN's documents . . . to be linked and accessible, especially given the history of so many failed documentation systems. Small but quantifiable steps seemed to be in order. Our first target, humble beginning that it was, would be the CERN telephone book. (Berners-Lee and Fischetti, 1999: 32)

The World Wide Web, therefore, entered CERN as an application that could be used to find CERN telephone numbers. Its designers, however, had a broader vision: if the end users could create links to their own content as well as content created by others, the different user groups could gain access not only to existing information but they would create a world of linked content that could become increasingly useful as more people started to use the system.

One critical technological challenge in creating such a system was to implement a mechanism that would make it possible to find the location of documents wherever they existed in the network. The KMS system used a master file to store information on the location of frames. Berners-Lee, however, made a different choice, which drastically reduced the complexity of his system. This, indeed, was perhaps the key innovation in the World Wide Web. He simply relied on the Internet host name resolution system. It provided a unified way to access computers using the Internet domain name system, thus using the existing Internet to avoid the need of a master file. The Internet domain name system was introduced in 1984 and it provided a way to access computers on the Internet using defined names, such as *info.cern.ch*. Thus existing procedures and computer programs for managing names in the global Internet were appropriated to solve a key challenge of the World Wide Web.

The World Wide Web, therefore, became possible for three reasons. There was content on the Internet and in the CERN information systems. The Internet technology had become available for many different operating systems, also outside the US Defense Department funded research sites, and advanced to the point where it provided standard services, which could be used to connect distributed computer systems without much programming. The NeXT computer had a rapid development environment with hypertext functionality. The World Wide Web didn't have to be novel to change the world. Existing technology and old ideas worked well because the world already had changed.

3.5 THE VISION OF XANADU

It is interesting to compare the World Wide Web to a very similar system that has been in development since the 1960s—a system that still waits for its breakthrough. This is Ted Nelson's Xanadu. In 1960 Nelson started to work on a global publication system which would allow anyone to publish digital content on a computer system, and which would facilitate reuse of already existing content (Nelson, 1995a: 31). The basic idea in Nelson's system was hypertext, a term that he coined in 1965. In the envisioned hypertext system, authors could link existing content to their own works. Nelson's goal was to produce a medium that would liberate individuals to become participants in the collective process of creating new interactive media. As Nelson himself describes:

Personally, I wanted a system for massively parallel creative work and study; more grandly, I sought to design the rightful literature and art canvas of the future, creating a technical, legal and commercial basis for a worldwide populist and participatory electronic literature of freely weaving screen transmedia—republishable and quotable without restriction—to the betterment of human understanding and freedom of expression and access. What better dream at 23? And still a good idea.   (Nelson, 1995*a*: 31)

A key concept in Nelson's system was that there were two different kinds of links: a node could link to another associated node, or it could include the content of that other node as a part of itself. An author could reuse content created by others by including it into her own work. Nelson planned to implement a copyright system that would automatically distribute royalties to the original authors whenever someone reused their material. Using a computer system even very small royalties could be accumulated. The system was to manage content integrity, versions, copyright, and royalty micro-payments.

When someone wanted to use a citation from someone else's work, she could simply put a link to the quoted work and the system would include it when it was browsed. Depending on the size of the quote, royalties would automatically be distributed to the author of the quote. Instead of copying the content, the quoted content was referenced using a universal resource name. The Xanadu system would then use this resource name to check where the actual quoted content could most easily be located. Nelson later coined the term *transclusion* to describe this mechanism for dynamically retrieving pieces of content (Nelson, 1995*a*). The system of intellectual property rights that made transclusion possible Nelson called *transcopyright* (Nelson, 1995*b*).

Nelson's Xanadu was an ambitious vision and he developed it for thirty-seven years before actually showing any software that implemented his ideas (Ditlea, 1998). It was based on a grand vision which required that the system would be implemented as a complete system. As many of the concepts in the system's design required functionality that was not available in current operating systems, Nelson and his colleagues had to develop the system from scratch. Just before Xanadu was supposed to be released in 1988, Nelson decided that it needed a complete rewrite.[3]

Xanadu developers have argued that Berners-Lee's WorldWideWeb has many fundamental limitations (Pam, 2000). For instance, WWW links use a universal resource locator (URL) that specifies the server machine which manages the referenced document. This means that there is no way to access the document if the server is down. Xanadu used a concept where each document had a unique name, but where documents could be stored in different machines, so that when a network

---

[3] More recently, Nelson has focused on ZigZag™, a multidimensional association space, which he has called Quantum Hyperspace (Nelson, 1998). An open source version, GZigZag, is being developed at the University of Jyväskylä (http://www.mit.jyu.fi/research/). Tuomas J. Lukka, who leads the developer team at Jyväskylä, describes ZigZag as a 'hyperstructure kit' (Lukka, 2001).

connection was broken or a server was down, the document could be retrieved from another location. Similarly, whereas much of the Xanadu architecture was based on solving the problem of fair reuse, the WWW didn't have any way to facilitate intellectual property management or compensation for reuse.[4] For example, as the links in the WWW were one-way links, there was no easy way for authors to know where documents and content were used. In contrast, Xanadu managed links in a way that made them visible both in source nodes and nodes where they were transcluded. The author, therefore, could easily monitor her author rights, as it was easy to know in what contexts content was reused. From the point of view of Xanadu developers, Berners-Lee's WorldWideWeb simply ignored the key challenges of building a world wide web of digital content.

3.6 SOURCES OF SUCCESS

In many ways the World Wide Web, therefore, was successful because its developers were able to rely on existing tools, technologies, and ideas. When the WorldWide-Web was distributed on the Internet in 1991, people outside CERN became able to use it as a platform for further development. In that process many new technologies were developed that addressed some of the concerns of the Xanadu visionaries. For example, the difficulty of finding documents that referenced a given node was alleviated by developing search engines, and intellectual property was managed by creating payment systems and secure Web servers.

Theoretically these solutions have often been inferior to the design envisioned by the developers of Xanadu and other sophisticated hypertext systems. At the same time the limitations of these solutions have provided many opportunities for improvement. Many innovators and entrepreneurs have swarmed in to solve acute problems generated in this process. At the same time large numbers of people have had to learn about the World Wide Web, its limitations, and what it one day could do.

More generally, as soon as a technological idea becomes a concrete technological artefact, its users can appropriate it in their own practices, and make it part of their own dreams. Although Xanadu tried to implement many radical ideas, and inspired the development of many successful commercial systems, so far it has not been implemented in a form where its visions could have been tested. The World Wide Web,

---

[4] Samuelson and Glushko (1993) compared Xanadu's intellectual property rights with existing copyright systems, noting that the Xanadu model was probably the most well-developed alternative to existing intellectual property systems. They also argued that the model had problems with its assumptions about user behaviour. Whereas Nelson had maintained that Xanadu operated according to existing copyright systems, Samuelson and Glushko noted that it is also different in many fundamental respects. They also noted that there was no 'fair use' in the Xanadu system as all use was supposed to be subject to royalties. Nelson, however, has argued that the motivation for the transcopyright system is to guarantee that reuse is possible (Nelson, 1995b). The traditional 'fair use' doctrine means that some use is free from royalties, whereas Nelson's concept of micro-payments implies that the cost can be small.

in contrast, emerged as a simple technological system, as a platform that could easily be appropriated for many different purposes. In this process, many early beliefs about the nature of this technology changed and many improvements were created. Abstract visions became redefined and the meaning of technology became concrete. After 1991, the World Wide Web was what it was used for.

Since the early works of Kondratieff and Schumpeter, economists have had an ongoing debate about the links between economic growth and innovation. Several authors have argued, for example, that there exist long waves in economic activity and that innovative activity intensifies when these long waves lead to a profound crisis in the economic system (cf. Freeman, Clark, and Soete, 1982; Perez, 1985; Mandel, 1995; Freeman, 2000). In this view, innovative activity clusters in periods where earlier technological paradigms become exhausted, and when people start to look for new ways out of the crisis. As the history of the World Wide Web shows, such cyclical theories miss an important point. The emergence of the World Wide Web creates a new space of innovation. This new space, in turn, creates new opportunities for exploration and makes innovative expansion possible. Instead of any obvious exhaustion of the capabilities of the dominant technological paradigm, there is a sudden opening up of a new land of riches. But many key contributions in this process come from outside the economic realm. Instead of gold, the primary driver is value and meaning.

When was the World Wide Web, then, invented? Was it invented when Nelson designed his Xanadu in the 1960s? Or was it, perhaps, invented when the file transfer protocol was taken into use in ARPANET, the predecessor of the Internet, in 1973? Or was it invented when Berners-Lee wrote his Enquire-program?

And who should we credit as the author of the World Wide Web? Was the real author Steve Jobs, who provided the NeXT machine that made WWW possible? Was it Tim Berners-Lee's father, who had the idea of programming computers so that they could make associations like the brain does? Or should the credit go to the Internet community, which created computer communication protocols that made it possible to move data between machines, independent of their operating systems, and which developed and maintained the domain name system that made universal resource identification possible?

The World Wide Web is essentially a combinatorial innovation, and therefore it is difficult to clearly define its author, or its date of birth. Here Nelson's vision becomes relevant again. The World Wide Web was created by 'transcluding' several existing resources and by organizing them in a new way. Many Internet-related innovations have a similar dynamic, as the following chapters show.

The traditional patent system, for example, is based on a different model of authorship. Therefore transcopyright and its improvements provide theoretically interesting alternative models for managing intellectual property rights in the new economy.

Richard Nelson has called such technologies 'cumulative systems technologies'. He argued that in such technologies strong patent protection 'may deter as many inventors as it encourages, and also adds inefficiency to the whole cumulative invention process' (Nelson, 1994: 2677). Nelson's comment was a response to a new intellectual property schema proposed by Samuelson *et al.* (Samuelson, Davis, Kapor, and Reichman, 1994; Davis, Samuelson, Kapor, and Reichman, 1996). Nelson agreed with Samuelson and her co-authors on the need to rethink intellectual property in software-related innovations. As the following chapters show, the problems noted by these authors are not necessarily limited to software, especially as software has become an important component in almost all technology.

Indeed, it is possible to argue that most of those innovations that underlie the so-called 'new economy' are based on recombination of resources. These resources, however, are not simply existing resources, as some economists tend to view them. Innovative recombination is not just about reshuffling existing pieces into new arrangements; instead, resources are created when the capabilities of the underlying networks of actors are taken into new uses. Recombination, therefore, is not a zero-sum game. Innovation is not only about reallocating resources, but also about changing the rules of existing games, and about creating new games where no one played before.

The World Wide Web itself is an example of this. In many ways, the World Wide Web continued to be invented after its first version was running on Berners-Lee's and Cailliau's desks. It avoided a quick death by transforming itself into an electronic telephone directory, and then became a platform for global information sharing. In 1993 the World Wide Web became Mosaic, and transformed itself in a couple of years into a dream of venture capitalists and Internet entrepreneurs (Reid, 1997). In the process 'World Wide Web' became ubiquitous. For example, instead of writing about the World Wide Web, newspapers simply started to post http-addresses in their articles, and started to call the World Wide Web 'the Web'. Simultaneously, the Web became a stock-market phenomenon and entered policy discussions. This can be seen, for example, from Fig. 3.1. The figure shows the

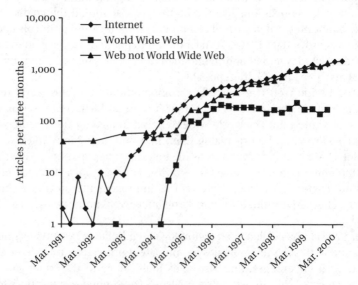

Fig. 3.1. How the 'World Wide Web' became 'the web'

number of articles in the *Washington Post* that mention the Internet, World Wide Web, or Web without explicitly mentioning World Wide Web.

As the history of the World Wide Web shows, the heroic innovation model misses many important parts of the story. It looks almost as if the anti-heroic model is right: if Berners-Lee had not invented the Web, someone else would have. Yet, when we try to understand how important new technologies emerge, it does not really matter who is the person that introduces a new innovation. A more interesting view is opened when we simultaneously contest both the heroic model and its traditional alternative. This requires that we rethink the concept of agency in technological change. To do this, we have to study in sufficient detail the various actors and events that interlink in the process of technological change. In the next chapter we focus on the actors and events that produced the Internet.

4

The Making of the Internet

In the previous chapter we saw that when the World Wide Web emerged, many of its key ideas were well known and had been implemented several times before. The breakthrough of the World Wide Web became possible to a large extent because the Internet, networked computers, modems, and digital content had become widely available. Whereas it took thirty-seven years for Ted Nelson to demonstrate his ideas of networked docuverse, Tim Berners-Lee programmed the first World Wide Web in about the same number of days. This was possible because his NeXT computer was intended to support hypertext applications and internet connectivity.

The key innovation in the World Wide Web, however, was in its underlying assumption: Berners-Lee knew that it would be impossible to predict how the users would use the system. In an industrial context, he might have had difficulties in convincing decision-makers that it is a great idea to build a system with unknown uses. In CERN's research context, it was somewhat easier to argue that the system was promising. Berners-Lee, for example, got his NeXT computer to study the possibilities of object-oriented computing. The legitimation for his unofficial project was based on the fact that it promised to help manage CERN's phone book.

Strictly speaking, even this key innovation in the World Wide Web, however, was not new. The developers of the NeXT personal computer were so convinced that graphical direct manipulation interfaces and hypertext were important that they programmed the required functionality into the operating system. The operating system developers didn't know how, exactly, this functionality would be used. They simply believed that someone might find it useful.

The Internet itself was developed in very much the same fashion. There have been many visions of the appropriate uses of the Internet during its history. Often these visions have been shown to be illusions. The Internet is today a system that binds together millions of users, all possible and impossible varieties of content, thousands of discussion groups, organizations and their business processes, and technology. This is possible because the underlying technical system is essentially a medium of communication.

In the course of its evolution, the Internet has overcome many perceived limits. People have improved it. Often these improvements have led to new opportunities that no one was able to predict, and which eventually made the Internet more than anyone dreamed of. The system didn't emerge overnight, and its history is replete with interesting examples of successes and failures of innovation. The Internet is a complex system of interlinked technologies that complement each other, providing a rapidly evolving platform that can adapt to new ways of doing things. A study on the evolution of the Internet therefore provides interesting and important insights on the nature of innovation. If we want to understand innovation, it is useful to dig deeper into the history of the Internet, to find out what its developers actually believed they were doing, and how the Internet eventually became what it is.

The key idea that makes the Internet possible is the digital packet-switched network. The idea itself was a modification of message switching that became important when telegraph traffic started to increase in the 1850s. Many of the important characteristics, benefits, and problems of the modern Internet can be found in the development of telegraphy. The key resources that made the global Internet possible were, however, created in the development of three competing technologies: telegraphy, telephony, and telex. To better see what is new in the Internet, it is useful to see what is not new.

In this chapter, I will show how the infrastructure was built that made the Internet possible. I will also describe the different goals, technical frameworks, and visions that produced the packet-switching technology. The focus in this chapter is on a very special phase of the history of Internet: the time when it was created, but when it did not yet exist as a technical system or artefact. In other words, I will describe a phase of development when the meaning of the Internet was defined by its developers, and when it had no users.

4.1 LAYING THE INFRASTRUCTURE

The Internet did not emerge from a vacuum. Many of the design choices that became important in the evolution of the Internet had their roots in much earlier systems. Indeed, some of the key concepts that underlie the Internet originate from the very first electronic communication and messaging systems.

When telegraph networks started to cover a large number of cities in the 1850s in the US, main telegraph offices became nodes that connected large numbers of telegraph lines in a single building. As the connections between telegraph offices become increasingly complex, the main offices evolved into sophisticated message switching centres. Messages coming from and going to nearby offices were handled on one floor of the building, and long-distance connections were operated from another floor. Pneumatic tubes were used to move messages within the office and between close-by offices. When messages arrived through the wire or the pneumatic tube, they were sorted on a table and forwarded using the building's own tube system. In 1875 the main office in London housed 450 telegraph instruments on three floors, linked by sixty-eight internal pneumatic tubes. The main office in

New York complemented its pneumatic tube system by 'check-girls' who delivered messages within its vast operating rooms (Standage, 1998).

When message traffic increased, it became increasingly attractive to solve the bottlenecks of telegraphy. One bottleneck was particularly obvious. Telegraphy required skilled operators who had learned Morse code. To address the problem, Charles Wheatstone developed the ABC telegraph, which he patented in 1858. It became known as the 'communicator'. The ABC telegraph was used for point-to-point communication in thousands of homes and offices. A similar system was developed by David Hughes, a professor of music in Kentucky. His system had a piano-like keyboard with black and white keys, each key representing a letter. The Hughes machine used a rotating disk of letters and a paper tape where letters were punched.

Both Wheatstone and Hughes telegraphs made it possible to send telegrams without an intervening skilled operator but their use was severely limited by the incompatibility of different systems. There was no way to connect them to the expanding network of Morse telegraphy.

To increase the efficiency on Morse telegraph, Wheatstone also developed an automatic sender that read Morse code from pre-punched paper tapes. This system, combined with existing Morse printers that recorded dots and dashes on paper tape, was able to send messages up to ten times faster than the best human operators. The Wheatstone Automatic was widely compared with the Jacquard loom that automatically wove cloth into patterns that were recorded on punched cards. Indeed, the Wheatstone Automatic was called 'the electric Jacquard'. The system made it possible for relatively unskilled workers to punch the tapes, and efficiently use the busiest lines also to send long messages, such as news reports. The Wheatstone Automatic was patented in 1858, and widely deployed after 1867 (Standage, 1998: 188–92).

Another obvious problem with telegraphy was that to connect to offices there had to be a wire between them. When traffic increased, a new line had to be installed. Furthermore, when one of the stations was using the line, if the other end wanted to send using that line, it had to wait until the line was free.

The first attempts to develop a system which could use one wire for simultaneously sending and receiving messages were already made by the early 1850s. It took, however, two decades before electrical theory had created sufficient understanding of the physics of telegraphy to make two-way telegraphy possible. The first two-way system was built and patented in 1872 by Joseph B. Stearns in Boston. To double the capacity of their lines, telegraph companies simply had to add the new duplex equipment to both ends of the line. The duplex was soon followed by the quadruplex, invented by Edison in 1874. In the same year, Jean Maurice Emile Baudot, working for the French telegraph administration, devised a telegraph that could use a single line to carry up to twelve lines' worth of traffic. The operators of the Baudot system used a five-key piano keyboard, and the messages were transmitted in five-bit binary form (Standage, 1998: 192–3).

As the theoretical understanding of telegraphy increased and it became clear that one line could carry more than one message at a time, several inventors tried to pack even more messages into a single line. One promising approach was the 'harmonic' telegraph. If a human being is put into a room with several Morse receivers, each tuned to a different pitch, it is easy for the human ear to single out one of the receivers, and regard the other sounds as noise. The idea of the harmonic telegraph was to use a series of reeds that vibrated at different frequencies. Messages were to be sent by stopping and starting the vibration of the reeds at the sending end, combining the electric signals generated by the reeds, sending this combined signal along the line, and again separating the 'sounds' at the receiver.

One of the inventors working with the harmonic telegraph was Alexander Graham Bell. On 2 June 1875 he was testing his equipment, and one of the reeds got stuck. His assistant, Thomas Watson, plucked the reed forcefully to free it, and Bell, who was listening at the other end of the wire, heard a complex sound that clearly resembled the twang of the reed. Bell realized that the system had the potential to transmit much more complex sounds than the inventors of harmonic telegraphy had assumed. At the beginning of 1876 Bell learned that Elisha Gray, who had already produced a working harmonic telegraph, was also devising a system that could transmit sound. Bell quickly filed a patent on 14 February 1876, and he was granted a patent on 3 March. The next week he made a breakthrough, transmitting speech for the first time over electric wire.

4.2 NETWORKING THE WORLD

During the nineteenth century electronic communication infrastructure was mainly based on telegraphy. Although the telephone quickly gained popularity, it remained a tool for short-distance communication until the beginning of the twentieth century. Long distances inevitably led to the fading of the signal. Under the best conditions, low-capacity lines could connect points that were 800 miles away. At the turn of the century, the application of theoretical models of transmission of electronic signals, based on Maxwell's equations, made it possible to design telecommunication lines that could transmit speech up to two thousand miles. The introduction of vacuum tube amplifiers in 1913 was the next major breakthrough, and the first transcontinental line was opened in 1915 between New York and San Francisco. Direct service from New York to Los Angeles opened in 1928. Before that, an average of 51 calls per day were routed via San Francisco. In January 1928 the *New York Times* reported that 'it is estimated that these circuits . . . will be used for an average of seventy-two conversations a day during the early Spring, which is the busy season for telephone calls between New York and Los Angeles' (quoted in Hugill, 1999: 70).

A major improvement in transmission occurred when twisted pairs of wire were replaced by coaxial cable. Coaxial cable was developed at the Bell Laboratories, in part to service the high-definition television systems being forecast all over the electrical world by the early 1930s. Using vacuum tube amplifiers and oscillators, speech was converted to a modulated signal, and several such modulated signals could be carried over the same line. The first coaxial cables were able to carry over 200 voice signals. At the end of 1939, in the US the coaxial long-distance system extended from the East Coast all the way to Cisco, Texas. In 1953 this L1 system was replaced by the L3 system that had transcontinental capacity of 1,860 voice circuits, or 600 voice circuits and one NTSC television signal (Hugill, 1999: 70).

As Hugill (1999) notes, for a long time the telephone was very much an American phenomenon. With few exceptions, telephony was developed by Bell's employees. To a large extent, this was the result of long geographic distances in the USA, but also its antimonopolistic policies that made it impossible for Bell to compete with telegraphy or postal services.

In Europe, telegraphy, telephony, and conventional mail were usually governed by a single state monopoly. In Britain, private telegraph companies were taken into public control and absorbed by the Post Office in 1869 (Standage, 1998: 172).

The telegraph also very early on became a tool for business in America. In Europe, the telegraph was more often used as a method of social and personal communication. Telegrams were used in Europe as an alternative to conventional letters. They remained the predominant form of rapid long-distance communication in European cities up to the 1950s and 1960s (Hugill, 1999: 77). In some cases, telegrams actually were never translated into Morse code. For example, in 1879 Paris announced a pricing system which made it possible to send cheap telegrams using the city's pneumatic tube system. Pre-paid telegram forms could be deposited into special mail boxes, handed over a telegraph counter in post offices, or put into boxes mounted on the backs of trams. At the end station of the tram the box was unloaded, and taken to a post office where the messages were distributed using the pneumatic tubes. The pre-paid forms became popular means for sending messages within the city, and were known as 'petits bleus' (Standage, 1998: 100).

Although European countries had national telephone networks, they had considerable difficulties in building international networks. Partly this was because there were no effective ways to coordinate international collaboration, and partly because telephone was much less used, and more expensive than in the US. Although an international committee, Comité Consultatif International des Communications Téléphoniques à Grande Distance (CCI), was set up in 1923 to agree on technical standards for international telephony, European post offices had no great interest in developing technologies for long-distance telephony.[1] International calling was expensive and impractical. For example, the average delay in setting up a call between Paris and Berlin was still over an hour in 1927. In contrast, AT&T claimed that its average delay for long-distance calls was 1.5 minutes in that same year. Also the cost of a 3-minute call was about four times more in France than it was in the US (Hugill, 1999: 79).

The US development was aggressively led by Bell Laboratories, which became a clear leader in communication technologies. Transcontinental lines required

---

[1] CCI later became CCITT, also known as Consultative Committee on International Telegraphy and Telephony. CCITT changed its name to ITU-T in 1993. It is now the telecommunications standardization sector of the International Telecommunications Union.

sophisticated technologies, and as Bell was unable to develop its business by moving to telegraphy and conventional mail, its way ahead was through expanding its long-distance business. Early on Bell was also forced to separate its equipment manufacturing from its telephony service business. This meant that its telephone manufacturing arm, Western Electric, could become a major supplier of telephony equipment to national telephone networks around the world. Western Electric set up its first factory outside the US in Antwerp in 1882, and by 1918 it controlled much of the world's telephone manufacturing business (Hugill, 1999: 57).

Centralization of telegraph services had some advantages. In Great Britain, for example, telegraph users could register a 'nickname' for their telegraphic address. Telegraphic addresses were assigned on a first-come, first-served basis, and a list of nicknames were published as a book. Over 35,000 telegraphic addresses had been registered by 1889, which generated a considerable income for the Post Office, since an annual charge was payable for each one (Standage, 1998: 172).

To a modern user of the Internet, it may be a rather astonishing fact that the first transatlantic telephone cable started to operate as recently as 1956. This was almost a century later than the first telegraph line was connected across the Atlantic in 1858. The first transatlantic cable, TAT-1, was built together by the American Bell and the British Post Office. It used one cable in each direction and provided 36 simultaneous channels. During the 1960s transatlantic capacity increased steadily, and the first transpacific coaxial cable was taken into use in 1964. Only in the 1970s, however, did intercontinental telephony become common. One important reason was the rapid reduction in costs. This can be seen in Fig. 4.1, which shows the investment costs of transatlantic telephone circuits divided by the minutes used. As the figure shows, the cost of transatlantic cable was more than one US dollar per minute until

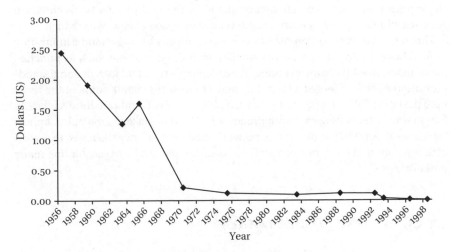

Fig. 4.1. Investment cost of transatlantic cable per minute of use

Source: FCC, 2000: table 12.

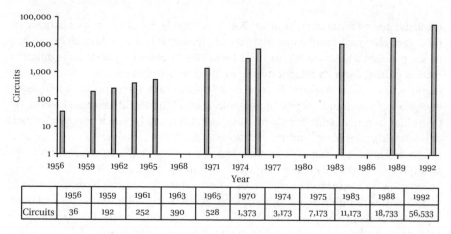

	1956	1959	1961	1963	1965	1970	1974	1975	1983	1988	1992
Circuits	36	192	252	390	528	1,373	3,173	7,173	11,173	18,733	56,533

Fig. 4.2. Total number of transatlantic voice circuits
Source: Hugill, 1999: 231, and FCC, 2000: table 12.

1970, after which it dropped rapidly. Whereas the estimated annual investment cost for one single transatlantic circuit in 1959 was over 200,000 US dollars, in 1970 it dropped to 18,773 dollars, dropped further to under 10,000 dollars in the 1980s, and to under 1,000 dollars in the mid-1990s (FCC, 2000: table 12).

Truly global communication networks emerged only after 1988, when the first fibre optic transatlantic cable was deployed (Hugill, 1999: 64, 231). The relatively recent development of international telephony can be seen in Fig. 4.2, which shows the number of transatlantic telephone circuits in use from 1956 to 1993. The rapid expansion of international communication capability is evident in the fact that there were only about 500 circuits available at the end of 1969 between North America and Europe. In 1992, in a single year, over 37,000 circuits were added.

This early history of communication networks already highlights some interesting points. Many innovations that we associate with the digital age, such as punched paper tapes, five-bit character codes, network congestion, and international standardization efforts, emerged a long time ago. On the other hand, some of the technological possibilities remained expensive, inconvenient, and rare for decades. As Fig. 4.1 shows, technological development is required to make technological opportunities real. Although transatlantic networks had existed since the 1850s, for over a century they remained out of reach to most people, and unavailable for many potential uses.

4.3 COMPETING TECHNOLOGIES

As the telephone became widely available for national and international use, and the costs of making a call dropped, the use of telegraphy decreased. The telegraph was partly replaced by telex machines. Telex allowed businesses to send short mes-sages to each other, without requiring skilled operators. Telex users could type their messages on a typewriter keyboard on paper tape, call the receiving telex machine, and send the recorded message from the tape. The receiving telex machine printed the message on paper, functioning effectively as a remote-controlled electric typewriter.

These three technologies were used in parallel for several decades. Teletype was first tested using Chicago & Alton Railroad telegraph lines, in 1908. After many improvements and slow growth in business, the head of the Associated Press became convinced in 1914 that teletype could be used to replace messenger boys who distributed news to New York newspapers. Within a year, all of the newspapers in New York City and nearby towns, as well as in Philadelphia, were receiving their press matter simultaneously from a transmitting set controlled by a single operator in the Associated Press office in New York City (Krum, 1925). The success of telex, therefore, was made possible by a novel use in a news service. Technological improvements, changing labour costs, and expansion in the production capability, however, also played prominent roles in the process. Howard Krum, one of the key innovators of telex, described the importance of learning by doing in his Morkrum company:

From the start good results were obtained, but as operation continued the inventors realized more and more that the operating requirements for commercial telegraph service were terri-bly exacting. The percentage of accuracy required was much higher than with any other form of mechanisms; it must work twenty-four hours a day; it must operate on good telegraph wires and on telegraph lines whose quality was impaired by rain and other adverse conditions . . . However, as in the case of the earlier installation, the inventors profited by their experience and went steadily along perfecting their apparatus, making changes here and there to improve its accuracy and to make it sturdier and simpler . . . the growth of the business was very slow. Telegraph companies and the railroads seemed loath to adopt the new system . . . However, the telegraph business continued to grow and good Morse operators became harder to secure, wages increased, and above all, the Morkrum system steadily improved . . . Due to increased business, Morkrum Company were able to enlarge their plant facilities, to engage expert assist-ants and to steadily improve their product.   (Krum, 1925)

According to Krum, already over 80 per cent of commercial telegrams were handled by printing telegraphs when he wrote his short history of the Morkrum company around 1925. Morkrum later changed its name to the Teletype Corporation. Teletype had a similar position to that of Bell in the telephone industry: it held many key patents to telex technology (Nelson, 1963).

In 1962 the revenue from international telephone calls exceeded for the first time the revenue from international telegraph in the US. Although the use of telex increased steadily until 1985, already in 1964 telephone became a bigger source of international revenue than telegraph and telex together. The revenues from inter-national services in the US are shown in Fig. 4.3. In 1995 telephony generated over 99 per cent of revenues from international services in the US. Telex revenues were 0.8 per cent and telegraph was more or less gone.

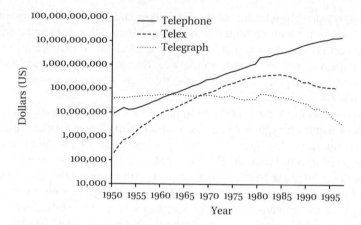

Fig. 4.3. Revenues from international services in the US, 1950–1997

Although telegraphy, telex, telephone, fax, and computer networks are alternative solutions to many communication needs, they are not equal. The 'entry barriers' of each of these technologies are different. Telegraphy required wire and batteries, and with these resources it opened the way for both telex and telephony. Telex connected organizations across the world and became a technology that enabled global financial flows. Telephony, in turn, provided the infrastructure for computer networks. Historically, each of these technologies has generated resources and behavioural change that prepared the world for the diffusion of more automated and more complex new technologies. This is one variation in the theme of path dependency. In the process of development, resources are accumulated and the direction of future change is partly determined by the characteristics and affordances of the accumulated resources.

When the development of the Internet is discussed, it is therefore important to remember that a major increase in global communications occurred only in the mid-1980s. Until the 1960s, even global corporations had quite independent regional subsidiaries. Although many multinational corporations existed before real-time communication was possible in practice, centrally managed multinational organizations emerged only in the 1960s when both jet airliners and communication networks could be used to keep company headquarters well informed about events around the globe (Hugill, 1999: 229). Ubiquitous global communication networks, however, were available only at the end of the 1980s.

The rapid expansion in global communications can be seen in Fig. 4.4. Until the 1970s, the percentage of revenues from international communications in the US was about 4 per cent, without much growth since the first transatlantic cable. In 1972 the revenue increased to 5 per cent, in 1988 it doubled to 10 per cent, and in 1997 it once again doubled to 20.5 per cent (FCC, 2000: table 1). The minutes used for international calls increased fivefold between 1987 and 1997 in the US.

It has often been noted that the world was in many ways tightly connected already by the end of the nineteenth century. International trade was widely spread

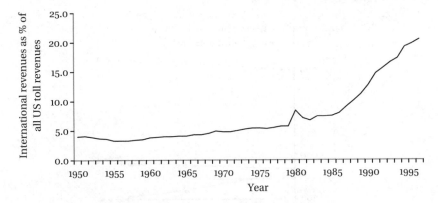

Fig. 4.4. **Total share of international revenues in the US, 1950–1997**

Source: FCC, 2000: table 1. The peak in the early 1980s is caused by the fact that Mexico and Canada were not counted as international traffic before that time.

before the First World War, scholars studied in foreign countries, and telegraph carried news from distant places with almost no delay. Qualitatively, globalization, however, changed during the last two decades of the twentieth century. It became a ubiquitous and unavoidable part of life in industrialized countries.

It is, however, also important to note that globalization was not in any simple way 'produced' by technological advances. Instead, global connections became ubiquitous as people and organizations appropriated their emerging potential for their own needs and purposes. Technological advances made it possible to reduce the cost of communication radically over several decades, and many uses that would have been impossible without these radically lower costs were invented in the process. The Internet itself is an example here.

Technological development and increasing communication thus feed each other, creating an economy of positive returns. The impact of this growth also propagates back to old methods of communication, as can be seen from Fig. 4.5, which shows the number of pieces of mail handled by the US Postal Service between 1980 and 1998. The total number of pieces of mail almost doubled between these years, from 106 billion to 198 billion. International mail switched from surface mail to airmail. Figure 4.5 shows only mail handled by the US Postal Service. It therefore undercounts the increase in international mail in the US, which is visible in the rapid growth of private mail services, such as Federal Express and DHL.

It has often been noted that the Internet was a 'research network' until the beginning of the 1990s. The history of communication networks shows that the fundamental reason for this was not the lack of new computer applications, such as email or the World Wide Web. Global communication infrastructure was being installed during several decades, without strong direct influence from computer networks, and after the mid-1980s this infrastructure started to make cheap international communications possible. Internet remained a 'research network' to a large extent because global networks didn't really exist before the end of the 1980s. As will be shown in detail below, the Internet was in many ways an impractical, speculative,

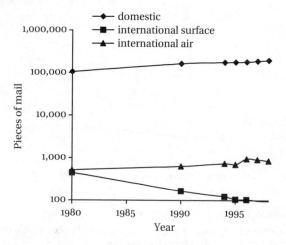

Fig. 4.5. The US Postal Service, pieces of mail handled
Source: US Census Bureau, Statistical Abstracts of the United States: 1999.

and geographically limited system when it emerged. Its benefits were often unexpected by its designers, and its underlying concepts were at many critical points evaluated by competent technologists who often remained sceptical about the practicality of the Internet. It is therefore quite a surprise that the Internet in just a few years became such a prominent aspect of everyday life.

It is an interesting characteristic of important new technologies that when they emerge, they are both new and old. When we retrospectively describe their emergence, they are new. They created a new world which, by definition, is something that didn't yet exist. At the same time, new technologies are almost always conceptually old. When they are actually taken into use, this is because there exist clear benefits for their use. Such easily perceivable benefits exist only if there are social practices that can readily apply and appropriate the new technology. This paradox means that important radical innovations are almost always invented 'after the fact', as unintended consequences of trying to solve a specific problem and finding unexpected opportunities to redefine problems and solutions. They are first introduced as solutions to specific problems, with clear understanding of their benefits, and in this process their meaning is changed and new benefits are discovered. In this way, the incremental innovations and improvements of the Morkrum Corporation radically changed the world of news reporting. At the same time, the meaning of printing telegraphs became redefined, and a new teletype industry emerged.

As will be seen below, it is exactly this process of reinvention and reinterpretation that made the Internet possible. But whereas new versions of telex remained similar to the original concept of the remote typewriter as long as telex was in use, the Internet has transformed itself several times during the few decades of its existence. This raises several interesting questions. Is the fundamental and ubiquitous impact of the Internet possible because it allows a new mode of technological development? Was the Internet the first important example of the dynamics of globally distributed innovation processes? Or does the flexibility of the Internet result from the fact that its most stable elements are not technical artefacts but standards and protocols? To answer these questions, we have to find out how the Internet itself was created.

4.4 MESSAGE-PACKETS AND RESILIENT NETWORKS: INNOVATION AT RAND

By the beginning of the 1960s it was generally understood that computers and communication technology were important for the future. The Cold War was turning into a technological competition between the Soviet Union and the US, and the risks of nuclear war were frequently publicly discussed. The launching of Sputnik in 1957 was a shock in the US, highlighting the possibility that the US could be left behind in technological competition. As a result, President Eisenhower appointed James A. Killian, President of MIT, as a presidential assistant for science. A new agency was set up in 1958 under the Department of Defense to oversee the development of the various US space programs. The new agency, the Advanced Research Projects Agency, ARPA, soon had its goal redefined, partly because the National Aeronautics and Space Administration (NASA) was established later in that year (Naughton, 2000: 78). The new goal was to prevent major technical surprises like Sputnik, and to coordinate and fund high-risk research and development projects that had a potentially high pay-off. Within a few years ARPA became a key driver in US computer technology (Norberg, 1996).

In the 1950s, advanced computer and communications networks were seen as a way to build systems that could provide early warning of nuclear attacks. Many early projects were extremely ambitious and failed to produce their intended results. In the process, however, many skills and ideas were developed that later on become important for the evolution of the Internet. The Whirlwind project, which started in 1947, developed many of the concepts of interactive computing, and its follow-up, the SAGE, was designed to coordinate radar stations and direct fighters to defend the US from an expected wave of Soviet nuclear bombers.

The RAND Corporation was one of the places where the interests of national security and technology met. It was founded in 1945 by the Air Force as Project RAND (derived from 'Research and Development'), under special contract to the Douglas Aircraft Company. Its goal was to provide scientific help to US military planners and to connect the military with research institutions and industry. In 1948 it was reorganized as a non-profit think-tank that focused on theoretical modelling of systems and strategies, and its name was changed to the RAND Corporation (RAND, 2000). Many strategic doctrines that underpinned the American stance during the Cold War originated at RAND (Naughton, 2000: 95, Abbate, 1999: 9).

According to Naughton (2000: 95), RAND was a surprisingly enlightened place that supported creative work. Partly this was made possible by the way it was funded. The Air Force awarded the RAND Corporation a grant once a year, leaving considerable freedom for the RAND managers to decide how to use the funds. Every week, RAND management circulated the letters that had been received from the Air Force and other federal agencies requesting help on various projects. If someone in RAND was interested, he or she could sign up for the project. If no one volunteered, RAND would send a form letter thanking the agency for its request but saying regretfully that the subject had been judged 'inappropriate' for RAND.

RAND, however, also conducted a number of major projects that were of acute interest to its sponsors. These 'support projects' were regarded as critical in preserving the goodwill of the Air Force, and legitimized the continuation of funding for RAND. According to Naughton:

It was a remarkably open, relaxed and productive regime which treated researchers like responsible adults—and, not surprisingly, elicited highly responsible and committed work from them.   (Naughton, 2000: 95–6)

In 1959 Paul Baran, then a 30-year-old engineer, joined RAND. Baran's background was in communications and remote measuring systems, but he was hired by the computer department. Soon after Baran entered RAND, he started to work with the problem of designing communications systems that could survive nuclear attack. As communication networks were becoming increasingly important both for early warning systems and maintaining the threat of retaliatory attack, the vulnerability of communications was becoming a major problem. Baran was specifically interested in developing a system that could guarantee that the president of the USA could launch retaliatory nuclear missiles even after most of the US communication systems were destroyed. This, according to Baran, would reduce the risk of nuclear war.

While working at the Hughes Aircraft Company in 1959, Baran had written a proposal for a communication network that could deliver short messages even when the network was partly destroyed. The idea was to 'flood' the network by sending a message through all available links as synchronized pulses. At each connecting node, the node examined all incoming messages, selected the one that came through the shortest route, sent it to all its neighbours, and threw away all corrupted or suspicious messages (Baran, 1964: v). Although this made reliable messaging possible, the system had a very limited capacity, and required that all nodes operated in synchrony.

The problem of survivability was close to the main interests of RAND. Frank Collbohm, President of RAND, had proposed in 1955 that the extensive network of US AM radio stations could be used to relay voice messages from one station to another, and other RAND projects had studied distributed communication networks (Baran, 1964: v). Baran's first proposal at RAND was based on showing that with some digital logic in the radio stations, it would be possible to get a simple message through the network in almost all circumstances. The idea was that every station tried to send the message until it started to receive copies of the message. The overlapping nature of the US broadcasting system would allow the message to diffuse rapidly through the surviving stations in the network. When the idea was presented to the US military, they were 'distinctly underwhelmed' (Naughton, 2000: 97). The feedback from the military was that they needed much higher bandwidth.

Baran went back to his drawing board, 'determined to give them so damn much communication capacity they won't know what in hell to do with it all' (quoted in Naughton, 2000: 97). He came up with two basic ideas. First, to guarantee that the network could operate even when many of its nodes were destroyed, the architecture of the network had to be distributed. There could be no single node that could be critical for the network, as the destruction of that specific node would destroy the net. Another key idea was that, to implement such a net, communications had to be digital. If the route that the signal took over the network could not be predicted, the only way to avoid the attenuation and distortion of the message was to convert it to a sequence of bits and transmit the signal in digital form. When the message was in digital form, additional bits could be added to it to make it possible to regenerate the message in every node it passed, thus making certain that the message was received in its original form. This meant that, instead of using AM radio stations as nodes, the nodes had to be digital computers.

Baran described the basic idea of distributed network in the following way:

The underlying concept of distributed networks is as old as man. Any interconnected grid of paths or roads may be considered as being a distributed network. When one drives to work over a distributed (or grid) road system and encounters a potential delay, it is possible to turn off, bypassing the traffic jam or obstruction. Thus, the actual route taken depends not only upon a predetermined route, but also upon the happenstance of encountering necessary detours which take us off the preferred shortest path. In spite of this uncertainty, and regardless of the number of detours, we almost always manage to get to work. On some mornings when we have a little extra time, we may chance to try a route that we have never taken before. If we find that this new route is quicker because of less traffic than our old route, we will probably take this newer route in the future. By this process, we learn in a relatively short time the quickest route between home and work. We may say that we have used a 'heuristic' process to learn a 'best' path in a network.   (Baran, 1964: v)

To route messages over the network, Baran proposed in 1960 a store-and-forward switching mechanism. He called this 'hot-potato routing'. The idea was that each node would try to get rid of its incoming messages as quickly as possible. The 'potatoes' were standard-sized digital packets with a header that contained the destination address of the packet. The nodes used the header to choose where to send the packet next.

In essence, this was the same store-and-forward mechanism that was used in telegraphy and conventional mail. The digitalization of messages, however, made a radical new idea possible. The message could be split into smaller blocks that could be sent separately, and reassembled by the receiver. The use of these 'message-blocks', as Baran called them, meant that the routing of messages could be done with simple machines and it became feasible to build networks with hundreds of nodes. Earlier proposals for message switching systems were based on the use of large computers. As the nodes were not able to know how much processing and storage incoming messages would require, they had to be ready to handle all possible situations. Baran's idea was that fixed-sized packets could be quickly moved on to the next node without much processing, so that the node computer didn't need expensive memory or complicated programs. As Baran developed this idea, he realized that standardized message blocks also made it possible that several users could use the same connections at the same time, and share the capacity of the line efficiently according to their needs. When there was little traffic on the line, someone could use most of the capacity, and when the traffic was heavier, the capacity was shared with the users without anyone being completely shut off from the network. Moreover, it was possible to build networks that connected links with different transmission capacity, as the switching station could send message blocks to the lines at the speed supported by the receiving line.

4.4.1 Sources of Innovation

In 1964, when Baran wrote a series of reports detailing his concepts, he also described the historical roots of the idea of distributed networks. He noted that 'The work of at least six separate disciplines is germane', but observed that 'these separate disciplines reside in different communities of interest' (Baran, 1964: v). According to Baran, relevant work on distributed networks had been done by those concerned with 'artificial intelligence', those concerned with communications within organisms and organizations, mathematicians working with optimization of flow in networks, mathematicians using dynamic programming to optimize incompletely understood and changing systems, those connected with civilian telephone switching, and, finally, military systems planners, especially those dissatisfied with existing network techniques. Baran noted that his report was written mainly from the viewpoint of the last group.

When Baran came up with these ideas, computers existed, but computer networks were rare. The idea of connecting several users to a single computer, which later become known as 'time-sharing', was widely discussed (Lee, 1992). In telephony, digitalization was just emerging. The first digital transmission system T1, was taken into commercial use in 1962 (Hugill, 1999: 71). The benefits of digital communications were well known since Claude Shannon had published his work on information theory in 1948 and 1949. Large computer systems had been built for national defence since the late 1940s. But, although many of the components of the digital computer networks were discussed, they were more ideas than everyday reality. Large computer systems, for example, were large in a very concrete sense: the computers in the Air Force's SAGE system weighted 250 tons each and had 55,000 vacuum tubes (Naughton, 2000: 69). In its early phases, Baran's design, therefore, relied on the availability of technology which didn't yet exist. Moreover, his ideas combined concepts that had been developed in different communities of interest, which used different languages, and didn't really know about each other's work.

When Baran discussed potential problems in implementing his network in 1964, he noted that the system was based on relatively cheap components that were rapidly becoming cheaper. According to Baran, the key challenge, however, was not technical:

We have discussed a new large communication system, one markedly different from the present in both concept and in equipment, and one which will mean a merging of two different technologies: computers and communications. People with competence in both these fields are not numerous. Our concern is whether we will have enough well-trained people capable of understanding both the communications and digital computer techniques to make this venture a success. Here may lie the real question of feasibility. Our present-day components are fully adequate. The difficult problems lie in hooking them together.   (Baran, 1964: XI, Postcript)

Baran's project was informed by the work of earlier researchers who had been trying to develop survivable and distributed communication networks. He was, however, also able to come up with his ideas for a very practical reason. He was assigned as a RAND representative to a high-level committee that was set up by the Pentagon to choose a communications system for handling Department of Defense records. The committee met every other week in Washington, so Baran was obliged to share his time between RAND offices in Santa Monica, California, and Washington, DC. As a member of the evaluation committee, he had the opportunity to ask dumb questions. As Baran himself noted: 'Dumb questions are only allowed without giving offense if you are a child of a member of a distinguished evaluation committee' (quoted in Naughton, 2000: 103). Baran was especially interested in understanding why the computers used in the message-switching system were so expensive and why they required large rooms. The answer, he discovered, was that the computers maintained detailed records of all their traffic they handled. He further concluded that 'the real reason every communications center office was built with that burdensome capacity was to be able to prove that lost traffic was someone else's fault' (quoted in Naughton, 2000: 104). Based on this observation, Baran used his extensive travel time to come up with an alternative design. RAND's travel policy supported this in an indirect way:

In those days RAND allowed its staff to travel first class if the trip was over two hours. This gave me the equivalent of one day per week to work in a comfortable setting. Almost all my writing was done either in airport waiting rooms or in airplanes.   (Baran 1990; quoted in Naughton, 2000: 104)

The first complete design for the network was ready by 1962. It was based on microwave links and 1,024 switching nodes, each about a shoe box in size. The distance between the nodes was kept short, about twenty miles, to keep radio transmitters and receivers simple. The system was designed to carry encrypted messages. Each node in the system could handle 128 secure telephone subscribers, together with 866 subscribers using computers and other digital devices (Naughton, 2000: 105).

The results of Baran's work were published in 1964. Following RAND's standard practices, Baran presented his work to various outside experts for comment while he was developing his ideas, and eleven volumes of reports that described Baran's work in detail were widely distributed. The first volume was also published in the March 1964 issue of *IEEE Transactions on Communications Systems*, and an abstract of that paper was published in *IEEE Spectrum*, with an estimated circulation of 160,000

(Abbate, 1999: 21). As Naughton observes, at first, the open distribution of such military-related work is rather surprising:

In retrospect, the freedom to publish secrets of this order seems astonishing. Just think: here were the results of advanced research on one of the most sensitive areas in US nuclear strategy—and they were placed in public domain. The US, in other words, just gave them away! They weren't even patented.   (Naughton, 2000: 105)

According to Baran, this openness made sense, however. The doctrine of retaliatory nuclear attack was based on balance of fear. Baran explained:

Not only would the US be safer with a survivable command and control system, the US would be even safer if the USSR also had a survivable command and control system as well! There was never any desire for classification of this work.   (Baran, 1990; quoted in Naughton, 2000: 105)

Although it may seem today that Baran's design had obvious advantages, proposals to go ahead with implementation were not successful. Baran was especially surprised to find that his strongest critics came from among senior technical experts at AT&T. One of the reasons why AT&T had difficulties in accepting Baran's ideas was that they implied that analogue telephony could not be developed into a reliable system. As all military telephone traffic went through AT&T lines, reputation for reliability was important. Furthermore, it seemed difficult to integrate this radical new system with existing networks.

Baran was, however, able to argue that further research and development would probably find ways to incrementally integrate the proposed network with existing systems. The Air Force staff accepted Baran's proposal enthusiastically. According to their normal procedures, an independent evaluation project was set up. The evaluation was very positive and recommended that the project should proceed. This never happened. As Naughton explains:

The reason was that it ran into the bureaucratic equivalent of a brick wall. Since 1949, successive US administrations had been trying to bring about the unification of the three branches of the armed services. Because of inter-service rivalries, it proved to be slow work. But in 1960 President Kennedy had installed as his Secretary of Defense a guy called Robert Macnamara,[2] a steely managerialist who had previously run the Ford Motor Corporation . . . By the time Baran's project reached the Pentagon, Macnamara's attempts at rationalization had produced a new organization called the Defense Communications Agency. This had been given overall responsibility for all long-distance communications for the three armed services. In the manner of these things, it was constituted on a modified Noah's Ark principle—a few of everything. It was headed, for example, by an Air Force general, and Army general and an admiral. More significantly, it was run mainly by ex-AT&T people who had zero or near-zero exposure to the new technology. 'If you were to talk (to them) about digital operation,' mused Baran, 'they would probably think it had something to do with using your fingers to press buttons.' . . . The problem as Baran saw it was that, if his RAND project were to go ahead, these boobies would be charged with building it.   (Naughton, 2000: 108)

---

[2] Naughton misspells McNamara here.

Baran discussed the issue with a friend who worked in the office of the Assistant Secretary of Defense and who was a key person in communications funding decisions. As a result of this discussion, Baran decided not to proceed with the project.

Paul Baran has often been described as 'the father of the Internet'. The current Internet implements many of the ideas that he developed in the early 1960s. When other researchers started to develop their ideas for distributed computer networks, Baran's work was quickly rediscovered. Yet, the Internet was not an obvious thing in 1964. As Baran himself noted, the system was difficult to explain and comprehend, a small network didn't really make sense, the required hardware did not exist, and people easily compared the proposed system with the existing telephone network, which for all practical purposes seemed to be working well, at least when no one had destroyed it (Baran, 1964: XI). Very few people had ever seen a computer. In the early 1960s every fifth household in the US didn't have access to a telephone, and in many states almost every second household was without a phone (US Census Bureau, 1999). As Abbate notes:

Many computer professionals have seen packet switching as having obvious technical advantages over alternative methods for transmitting data, and they have tended to treat its widespread adoption as a natural result of these advantages. In fact, however, the success of packet switching was not a sure thing, and for many years there was no consensus on what its defining characteristics were, what advantages it offered, or how it should be implemented—in part because computer scientists evaluated it in ideological as well as technical terms. Before packet switching could achieve legitimacy in the eyes of data communications practitioners, its proponents had to prove that it would work by building demonstration networks. The wide disparity in the outcomes of these early experiments with packet switching demonstrates that the concept could be realized in very different ways, and that, far from being a straightforward matter of a superior technology's winning out, the 'success' of packet switching depended greatly on how it was implemented.   (Abbate, 1999: 7–8)

4.5 TIME-SHARING AND NETWORK SOCIETY: WORK AT NPL

Baran's work was very much based on his ability to evaluate existing proposals for message-switching and to learn that they were based on assumptions that were not relevant for solving the fundamental problem at hand. He also made the assumption that although suitable technology didn't exist for the network nodes, it could be developed relatively easily from available components. Baran had access to competent experts in the area of communication networks, he had a vision of the basic architectural requirements, and RAND gave him time to develop his ideas for several years. The resulting series of reports is, indeed, a careful, insightful, and clearly written study on the benefits of distributed digital networks. Yet, Baran's proposals were not implemented. In many ways he was a prime example of the traditional heroic innovator. He was trying to develop new technology as an individual innovator. His failure actually to implement the system may result from the nature of the technology he designed. Distributed communication networks, after all, are distributed networks. Networks are difficult to set up locally, and Baran also had to convince many existing communities that they should change their practices.

While Baran was trying to find someone to implement his plans, another attempt to develop digital computer networks was launched in the UK. It had quite different goals from Baran's network, but it was based on strikingly similar technical ideas.

In 1963 Harold Wilson had made technology one of the key issues in UK politics, calling on labour and management to join in revitalizing British industry (Abbate, 1999: 22). When Labour came to power in the 1964 election, Wilson quickly started to implement several projects that tried to promote new high-tech industries and to stop the perceived 'brain drain' of world-class scientists. Wilson believed that the British computer industry would be destroyed by competition from the United States unless the government intervened urgently. As a result, funding for computer research and commercialization of new technologies increased substantially.

The basic concept of digital packet networks was invented independently of Baran's work in 1965 by Donald Davies, a British scientist working in the National Physical Laboratory. The NPL had become one of the places that tried to close the 'technology gap' between the United Kingdom and the United States.

Donald Davies had joined NPL in 1947. He worked with the early British computer initiative ACE, which was to be the answer to the American challenge on computer technology, and did some work including a classified communications project which involved sending data securely over telex links from a weapons-testing range to a processing centre (Naughton, 2000: 121–2). Davies also developed a computer program to simulate road traffic (*The Times*, 2000). When the Labour administration launched its Advanced Computer Technology project, Davies was nominated to head it in 1963. As a result, Davies had a good understanding of the developments in computer technology. In May 1965 Davies went on an extended visit to the USA. The 'ostensible excuse was to attend the International Federation for Information Processing (IFIP) Congress in California, but because there was always pressure in the NPL to justify the expenditure of a transatlantic trip, Davies set up a number of visits to various research centres' (Naughton, 2000: 122). He went to Dartmouth, New Hampshire, where a new computer language BASIC was being developed, and visited MIT and RAND, to learn about interactive time-sharing computing. After the trip, he organized a three-day seminar on time-sharing, which was followed by an open one-day meeting organized by the British Computer Society.

Time-sharing was becoming a hot topic in computing as it allowed several users to connect to a single computer using terminals, and use the machine simultaneously. In a time-shared system, the processor time was split among independent users, who each had control of the machine for short periods of time (cf. Lee, 1992). The speed of the processor made it possible to switch from one user to another so fast that in effect everyone thought they had the complete machine in their own use. This allowed cost-effective real-time access to large computers and made many new interactive uses of computers possible (Fano, 1963).

At the same time, it was possible to imagine that some of the terminals could be remote terminals. If data communication could be possible through public telephone networks, expensive computers could be used from distant locations. Indeed, in the mid-1960s, research centres used time-sharing regularly and businesses began to offer commercial time-sharing services to customers who would rent or buy a terminal, connect to the service using a modem, and pay for the service at a hourly rate (Abbate, 1999: 25).

Telephone systems, however, were not well suited for computer communications. They operated on the principle that a circuit had to be connected between the callers. In the 1960s this was still often done by human operators. But even the automatic direct dialling systems assumed that a circuit had to be reserved for every call. Telephone networks were fundamentally different from telegraph networks in this sense. Whereas the latter operated in 'store-and-forward' mode where a message jumped from one node to another, without requiring continuous connection between the end points, telephony was based on opening a connection between the endpoints and keeping the connection reserved until the call ended.[3] This concept was not well suited for computer communications. Computers needed to send bits over the line only infrequently, when the user pressed a key on the keyboard, or when the computer sent characters back to the user. Moreover, the time that was needed to set up a connection was very slow. Therefore it was also impossible to expect that a terminal and the computer could open the line only when they needed it.

A few days after he had organized the seminar on time-sharing systems, Davies came up with an idea that would solve this problem. If messages were split into small blocks, a store-and-forward network could be a great way to connect interactive computer users to a central processor (Davies, 1982). Davies estimated the number of messages that a large city would generate, and noted that if the nodes were to forward the messages fast enough they could easily handle a million users (Davies, 1965).

In November Davies wrote two notes on his ideas and sent them to a number of people 'who in their replies showed some interest' (Davies, 1982). A month later Davies wrote a more detailed proposal, entitled 'Proposal for the Development of a National Communication Service for On-Line Data Processing' (Davies, 1965). He envisioned the network as offering many different services for businesses and recreational users. The list of potential uses clearly shows the different goals of Davies's concepts and Baran's concern with survivable communications. According to Davies (1965), the services provided by the network included:

- numerical computation at various levels of generality
- editing and typesetting of text
- design services and problem oriented languages
- availability of goods for sale

[3] In automated direct dialling, the connection was opened by sending requests from one switch to another using a store-and-forward mechanism. When the switchers found a route from the caller to the receiver, a circuit was opened and the connection become a fixed connection until the end of the call. Direct dialling telephone networks, therefore, combined direct 'real-time' links between callers and asynchronous dispatching of messages between the switches.

- ordering of goods
- invoicing, delivery notes, etc.
- booking of transport
- banking, establishing credit
- remote access to national records, e.g. MPNI, tax, police, medical, on a secure basis
- betting.

Davies saw that the connection of users to computer services was of primary importance, but he also noted that the network could be used for people-to-people communication, and could replace telex. Davies pointed out that the system could also be used for machine-to-machine communication, for example, for road traffic control, monitoring and controlling of utility services, pipelines and automatic meteorology stations, burglar alarms and other security devices, as well as for controlling of telephone switching. To test the concepts, he proposed that a pilot network should be set up in central London. Davies circulated this proposal widely among banks and the UK Post Office, who had the monopoly on communications networks. According to Davies (1982), 'the reaction was generally appreciative but not enthusiastic'.

In March 1966 Davies gave a public lecture at the NPL on 'The Future Digital Communication Network'. Over 100 people attended the lecture of whom eighteen were from the UK Post Office. After the lecture he was approached by a man from the Ministry of Defence, who told him that some remarkably similar research had been done in the US by Paul Baran (Davies, 1982; Abbate, 1999: 29; Naughton, 2000: 126).

To distinguish the messages, as they were seen by the user, and the blocks of data that traversed the network, Davies decided to call the latter 'packets'. Each packet consisted of the data it carried and a header that contained information on the packet's source and destination. The header also contained a check digit that could be used to make sure that the packet arrived without errors, and a sequence number that allowed the receiver to assemble the incoming packets back to their original sequence and retrieve the message that was sent. The maximum size of a packet was 128 bytes.

In the summer of 1966 Davies wrote a twenty-five-page 'Proposal for a Digital Communication Network'. The proposal described a design of his packet-switched network, and estimated costs of building such a network. Davies's plan also included 'interface computers' through which remote users could connect to the network. Although Davies designed some redundancy to his network, his main concern was not in building a network that remained operational after a nuclear strike; instead, his main concern was to provide interactive access to large numbers of users, who could not otherwise access expensive computers. Packet switching was for Davies a way to share communication lines effectively, in a similar way to that in which time-sharing efficiently divided the resources of the central processor. Davies also saw commercial potential in packet-switching technology, and believed that it could directly contribute to Harold Wilson's plan to revitalize the British economy (Abbate, 1999: 27). Indeed, already in his first proposal six months earlier, Davies had noted that it was important to start the development work early to influence the emerging international standards, and that 'it is very important not to find ourselves forced to buy computers and software for these systems from USA. We could, by starting early enough, develop export markets' (Davies, 1965).

The network that Davies proposed was national in scope, but the problem was that national networks were a monopoly controlled by the General Post Office. Its managers had little interest in building a nationwide data communications infrastructure. This was quite understandable: there were not many users or uses for such a network anyway. According to Davies, the Post Office reaction to his March 1966 lecture was surprisingly tolerant. People in the Ministry of Technology were also interested. The proposal, however, 'was never considered seriously as a practical possibility' (Davies, 1982). Later, Davies noted the basic problem with the GPO:

I had been in contact with them enough to know they were a pretty large, monolithic organization, in which to get anything done you have to convince a lot of departments. The fact that my ideas had had any impression on them at all was to me rather amazing, and I didn't expect them to put a lot of effort into it quickly.   (Davies, 1986; quoted in Abbate, 1999: 225)

As Davies felt that it was difficult to convince the managers of the GPO, he decided to propose a small-scale system that could demonstrate the concepts in real life. During the spring of 1966 he had been discussing this possibility with his colleague Derek Barber. The benefit of this approach was that an internal NPL project could be started without many formalities. As Davies noted: 'In those days we had a certain amount of flexibility in our research programme enabling us to begin such a project without undue formality' (Davies, 1982). In July 1966 he proposed that NPL should build its own network that could connect its computers and show the concepts of packet networks in actual use. The network, named 'Mark I', would serve as a demonstration of packet switching, advance the state of knowledge in the field, and support the operational computing needs of the NPL's scientific and administrative personnel (Abbate, 1999: 29). In August, Davies was promoted to head NPL's computing division (at that time known as the 'autonomics division'), which gave him resources to proceed with the plan. The project started in 1967. The development team was headed by Derek Barber. Roger Scantlebury was the technical leader, Keith Bartlett was responsible for the hardware development, and Peter Wilkinson was in charge of software development. By early 1967 the team had a plan to link ten computers, ten high-speed peripheral devices, fifty slow peripherals and graph plotters, six computer terminals and forty teletype terminals. By July 1967 Davies had detailed enough plans so that he could request funds from the Lab's Steering Committee to start building the system (Naughton, 2000: 129).[4]

The architecture of the Mark I network became fundamentally limited due to budgetary limitations. The original plan was to build a demonstration network with three computer nodes, but limited funds meant that Davies had to start with only

---

[4] Naughton notes that the team was lead by Scantlebury. Scantlebury was the technical leader, and Barber was the head of the team.

one node computer. Davies didn't think this would be a major limitation, however, as a multi-node network could be simulated by a computer program, and its routing problems could be studied without actually building a multi-node network. The original idea was to support connections over a wide area; in practice, however, the architecture of the network became what is today known as a local area network.

Politics and funding played a considerable role in the development of Mark I. The English computer manufacturer Plessey had designed a minicomputer specifically for data communications, and the NPL team decided to plan their network using it as the node computer. The British Ministry of Technology was, however, actively trying to rationalize the British computer industry to improve its competitiveness. The Ministry promoted the idea that computer manufacturers should limit the number of different computer models to gain scale benefits in manufacturing. As a result, Plessey abruptly withdrew the computer that the NPL team was planning to use, and the NPL team had to start from scratch. As Abbate notes:

Bowing to this policy, the Plessey Corporation canceled its plan to produce the minicomputer that the NPL team had chosen for its network interface. This delayed the NPL project and forced the NPL designers to make up for the lost functionality of the Plessey computer by increasing the complexity of other parts of the system.   (Abbate, 1999: 34)

The limited funding and high cost of computer equipment meant that the NPL team could proceed only slowly. The project started in 1967, the first network interface minicomputer was installed in 1969, and the local network at NPL began operation in 1971. The software was subsequently rewritten, and the full use of the improved Mark II network started in 1973. The Mark II remained in active service in NPL until 1986.

The work at NPL did, however, have considerable influence on the development of computer networks. Already in December 1967 Davies was invited to a Consultative Committee for International Telephony and Telegraphy (CCITT) meeting in Geneva, where data communication was discussed by a group of telecommunications experts. Fred Warden, from IBM, gave a long lecture on the future of data communication from the industry point of view. The representative of the UK Post Office, who was chairing the meeting and who knew about Davies's proposals, asked him during a break to give a short presentation on the principles of packet switching. 'I was thus able to introduce packet switching to CCITT', Davies commented later. 'At that meeting a resolution was passed for the establishment of a Joint Study Group on New Data Networks in the 1968–72 Session of CCITT' (Davies, 1982). Subsequently, the members of the NPL group played an important role when the packet switched data networks were standardized by European and international telecommunications authorities. The development was, however, slowed down due to the varying interests of the different parties. For example, early on Germany tried to promote data networks that were based on an extension of their telex system. The German proposal was based on a new kind of switch that Siemens had developed (Davies, 1970).

Eventually, the CCITT work led to the X.25 standard for data communications. This standard was developed in a hurry when there started to be increasing fears in

1975 that IBM could gain monopoly in data communications using its proprietary standards. In part, the tight schedule for CCITT resulted from the fact that its standards had to voted by the entire CCITT membership, which met only every four years, in the CCITT's plenary (Abbate, 1999: 154). The X.25 was accepted as a CCITT recommendation in 1976. X.25 networks were widely deployed by the national PTTs, and they were the main competitors for the current Internet networks until the mid-1990s.

The Wilson government's interest in computing technology also meant that there was increasing oversight and intervention. As Abbate notes:

The politics of the day and the culture of some British institutions hampered Davies's ability to implement his ideas and fulfill his aim of keeping the United Kingdom ahead of the United States in computer networking. In the late 1950s the NPL had been oriented toward pure research, but under the Wilson government there was a marked increase in government oversight and intervention.   (Abbate, 1999: 34)

Abbate goes on to quote a NPL scientist:

'Schemes for improving the service given to the nation were constantly being hawked from above ... Open-ended research was severely cut back and in its place all research projects had to have a 'customer,' who had to be persuaded of the viability and value of each project and agree to make available the funds to carry it out . . . Meetings required regular preparations of cases by Laboratory scientists in time which could ill be spared from practical work'.   (quoted in Abbate 1999: 34)

Political interests played an important role in the development of UK networks throughout the 1970s. In 1970 Larry Roberts, head of the US ARPANET project, proposed that ARPANET and the NPL network should be connected. According to Peter Kirstein, 'the timing could not have been worse from the British perspective'.

The problem was that the British government had just applied to join the European Community; this made Europe good and the United States bad from a governmental policy standpoint. NPL was under the Department of Technology, and Davies was quite unable to take up Robert's offer. He had to concentrate on European initiatives like the European Information Networks.   (Kirstein, 1999: 39)

As Kirstein, from the University College of London (UCL), had been interested in the US Arpanet project already since its beginning, it was decided that UCL would become the first ARPANET node outside the US.

4.6 INTERACTIVE COMPUTING: AUGMENTING THE HUMAN MIND

Donald Davies died on 28 May 2000. A few days later *The Times* published an obituary. It noted:

After working with Alan Turing, the scientific genius who first conceptualized computer programming, Donald Davies went on to make one of the crucial breakthroughs that made possible modern computer communications. He pioneered packet-switching, which enables the exchange of information between computers, without which the Internet could not function. (*The Times*, 2000)

Most histories of the Internet, however, argue that the Internet originated in ARPA, and that its origin can be traced back to J. C. R. Licklider. In 1960 Licklider published an article on 'Man–Computer Symbiosis', which outlined his vision on computing (Licklider, 1960). His article introduced several themes that subsequently played an important role in the development of the Internet.

In the 1940s, early users of computers saw them mainly as fast programmable calculators. By the mid-1950s, computer researchers were becoming increasingly interested in automated decision-making. Instead of solving differential equations, computer scientists were increasingly using computers to uncover the secrets of the human mind. When the first conference on artificial intelligence (AI) was organized in Dartmouth in 1958, AI had become almost equivalent to 'computer research'. Many prominent researchers were excited about the challenges and possibilities of programming computers to process symbols and thoughts, instead of numbers (McCorduck, 1979).

When Licklider published his paper on man–computer symbiosis, he highlighted the possibility of a third way. AI researchers were focusing on the automation of thought. Licklider, however, argued that, at least for the time being, a more promising possibility was to develop computers as tools that could help their users to think. Licklider had conducted an informal time-and-motion study in 1957, using himself as the subject, to better understand what mental work actually consisted of.[5] He found out that there was very little 'thinking' in technical research and engineering work. Most of the time went to finding information, plotting of graphs, instructing an assistant to plot graphs, and getting into a position to think:

Throughout the period I examined, in short, my 'thinking' time was devoted mainly to activities that were essentially clerical or mechanical: searching, calculating, plotting, transforming, determining the logical or dynamic consequences of a set of assumptions or hypotheses, preparing the way for a decision or an insight. Moreover, my choices of what to attempt and what not to attempt were determined to an embarrassingly great extent by considerations of clerical feasibility, not intellectual capability.   (Licklider, 1960)

Licklider argued that humans and computers had different characteristics and that therefore they could effectively complement each other.[6] A computer could be used as an aid in formulating problems interactively with the user, and to support thinking in 'real time'. This, however, required development of new computer languages, user interfaces, and new ways to store information in the computer memory.

---

[5] This was probably inspired by a similar study done earlier by Wiener, cf. Naughton, 2000: 67.

[6] This idea, in itself, was not an exceptionally radical one. The idea of man–machine symbiosis was quite popular in the early 1960s. There was a lot of interest in, and funding for bionics, which involved the interplay between biology and technology. For example, in the 1950s Ross Ashby had proposed 'intelligence amplifiers'. The only form of bionics that remained popular after the 1960s was the AI based on symbol processing (Rosen, 2000: 283–96). The current interest in augmentation systems picks up many themes that were popular in the 1950s and 1960s.

Licklider argued that pen-based input and desk-surface display were necessary and that limited speech recognition was becoming possible in the next few years. He also proposed that computer output could be projected on a wall when a team of people needed to work on a task together. As computers were much faster than humans, and as it was inconceivable that expensive computers could be used by only a single user, time-sharing was necessary. Licklider suggested that maybe in ten or fifteen years, there could be 'thinking centers' that could incorporate the functions of libraries together with the electronic storage and retrieval systems that he proposed:

The picture readily enlarges itself into a network of such centers, connected to one another by wide-band communication lines and to individual users by leased-wire services. In such a system, the speed of the computers would be balanced, and the cost of the gigantic memories and the sophisticated programs would be divided by the number of the users. (Licklider, 1960)

Licklider continued working with these ideas, and developed later a vision of computing which he called the 'Inter-Galactic Network'. When Licklider became the head of ARPA's Command and Control division in 1962, he quickly started to sponsor a number of projects that developed time-sharing computers and interactive computing. These led naturally to research on computer networks.

Licklider's vision was to a large extent based on his first-hand experience with computers and interactive computing, and in his background in human sciences. He had earned undergraduate degrees from Washington University in physics, mathematics, and psychology, and in 1942 he received a Ph.D. in psychology from the University of Rochester. During the Second World War he was a research associate and research fellow in the Psycho-Acoustic Laboratory at Harvard University, working on defence projects. After the war he stayed at Harvard until 1950, when he joined the Psychology Department at MIT as an associate professor.

During his Harvard years, Licklider also joined a discussion group that was organized by Norbert Wiener. Wiener had been a member of a similar group of researchers in the 1930s, led by neurophysiologist Arturo Rosenbleuth, where mathematical models of biological systems were actively discussed. During the war, Wiener had worked with the problem of automatic control of guns for air defence, and he realized that the human neuro-motoric system had many similarities with automatic targeting systems. He summarized these ideas in his book on Cybernetics, which was published in 1948. It quickly became one of the most influential books in the latter half of the twentieth century. In the 1950s cybernetics and systems theory were discussed with great enthusiasm in the US, Europe, and the Soviet Union. These ideas played an important role in both setting up the RAND and, for example, in the design of the US SAGE air defence system. Wiener set up his own circle to discuss the issues related to cybernetics and modelling of biological systems in 1948 (Naughton, 2000: 55–66).[7]

---

[7] Cybernetics emerged from a very interdisciplinary collaboration with many brilliant researchers, including Nicolas Rashevsky, Warren McCulloch, Walter Pitts, Margaret Mead, Arturo

Licklider was greatly impressed with cybernetic ideas and when he joined MIT, he got access to the most advanced computer systems available at that time. A year after he left Harvard for MIT, MIT set up the Lincoln Laboratory as a facility that specialized in air defence research. Licklider became the head of the laboratory's human engineering group. During this time he learned to program TX-2, one of the first computers that used transistors, and learned about the $61 billion, 7,000 man-year SAGE project.[8] SAGE was the first major system of interactive and networked computing. Its scope is reflected in the fact that its phone bill was millions of dollars a month (Naughton, 2000: 70).

In 1957 Licklider became a vice-president of Bolt, Beranek and Newman (BBN). BBN, which later become a key contractor in the ARPANET project, was at that time a company engaged in studies of acoustics, psychoacoustics, human—machine systems, and information systems. Licklider's paper on Man–Computer Symbiosis was partly based on the work performed by a small research team that he had organized and managed at BBN (Taylor, 1990). During the 1950s, Licklider also consulted local laboratories and companies, served on the Air Force Scientific Advisory Board, and consulted directly with the Department of Defense (Norberg, 1996: 44).

In 1962 ARPA was looking for a manager to head a new program on behavioural science, and to manage the time-sharing activities in the System Development Corporation in California. The director of ARPA, Jack Ruina, found two candidates. After some discussions, Licklider was asked to join ARPA to head the Command and Control Division and to set up a new division for behavioural sciences. By skilfully using his somewhat ambiguous job description, Licklider focused on interactive computing. As Naughton notes:

Lick arrived at the Agency on 1 October 1962 and hit the ground running. He had two over-riding convictions. The first was that time-sharing was the key computing technology; the second was that the best way to make progress in research was to find the smartest computer scientists in the country and fund them to do whatever they wanted.   (Naughton, 2000: 80)

Licklider's research program soon changed its name to Information Processing Techniques Office (IPTO), and became the biggest sponsor of computer research in the US. Although it was one of the smaller programs within ARPA, its annual budget was greater than the total amount of money allocated to computer research by all

Rosenblueth, Oskar Morgenstern, Gregory Bateson, and others. An early seminar at the Institute for Advanced Study at Princeton in 1942 brought together mathematicians, physiologists, and mechanical and electrical engineers. It was a great success, and a series of ten seminars was arranged by the Josiah Macy Foundation. This led to development of cybernetics, general systems theory, and systems dynamics (cf. de Rosnay, 1979).

[8] Jay W. Forrester, who later founded Industrial Dynamics and developed influential system dynamics models for the Club of Rome, was the head of the Lincoln Laboratory, and headed the development of the SAGE (Semiautomatic Ground Environment) system from 1952 (Astrahan and Jacobs, 1983; de Rosnay, 1979). The first SAGE computers used vacuum tubes. Lincoln Laboratory also developed a series of transistorized computers. The first of these, the experimental TX-0, started to operate in 1956, under the supervision of Kenneth H. Olsen. Olsen later founded the Digital Equipment Corporation, commercializing ideas that originated in the TX computer architecture (Ceruzzi, 1998: 127).

government-supported agencies. The funding from IPTO also launched the first computer science Ph.D. programs in the US, in UC Berkeley, CMU, MIT, and Stanford (Taylor, 1990).

4.6.1 Memex and the oNLine System

A few months after joining ARPA, Licklider was able to arrange funding for Douglas Engelbart, who was developing interactive computing at Stanford Research Institute (SRI). Engelbart and Licklider shared the basic vision of computers as tools to augment human thinking. It is impossible to understand the evolution of the Internet without knowing Engelbart's contributions. But, as always, visionaries do not emerge from a vacuum. Engelbart was greatly influenced by an article that he read in 1945. This was Vannevar Bush's article where he described a vision of the 'memex' system.

Bush's article was published in the *Atlantic Monthly* in July 1945 and it became influential partly because it was written by the '*de facto* science advisor' of President Roosevelt (NSF, 2000: 1–4). The article was widely quoted, and a condensed version of the essay was published in *Life*, September 1945. Bush had written the first full draft of the article already in 1939 (Naughton, 2000: 212–15).

Bush's memex was based on his conviction that the human mind operates through association, and that existing methods of information processing had become inadequate with the increasing amounts of information. Bush had joined the Electrical Engineering Department at MIT in 1919 and stayed there for twenty-five years, eventually becoming Vice-President of MIT. During this time he developed an analogue computer, the Differential Analyzer, and optical and photo-composition devices which rapidly selected items from banks of microfilm. Bush left MIT in 1939 to become President of the Carnegie Institution and during the war he was recruited by President Eisenhower to run the scientific defence efforts (Naughton, 2000: 52).

In his *Atlantic Monthly* article, Bush argued that information retrieval should be supported by mechanized machines:

Professionally our methods of transmitting and reviewing the results of research are generations old and by now are totally inadequate for their purpose . . . The difficulty seems to be, not so much that we publish unduly in view of the extent and variety of present day interests, but rather that publication has been extended far beyond our present ability to make real use of the record. The summation of human experience is being expanded at a prodigious rate, and the means we use for threading through the consequent maze to the momentarily important item is the same as was used in the days of square-rigged ships. (Bush, 1945)

To overcome the problem of retrieving relevant information, Bush proposed a machine that used a microfilm recording system, a keyboard, and a set of buttons and levers. The machine was built into a desk, on top of which there were slanted translucent screens that could be used to project the content of microfilms. Such a 'memex' was to be a device 'in which an individual stores all his books, records, and communications, and which is mechanized so that it may be consulted with exceeding speed and flexibility. It is an enlarged intimate supplement to his memory' (Bush, 1945).

A key to memex was that each microfilm document had a space where the user could record links between the various documents. Recording such associations, the user was able to build 'traces' of documents, and move rapidly along the traces, creating new ones while doing this. Bush estimated that using an improved microfilm system, the memex could have a very large storage, allowing the user to store 5,000 pages of material a day for hundreds of years.

Bush envisioned a hypertext system that closely resembles Berners-Lee's original vision of the World Wide Web:

The owner of the memex, let us say, is interested in the origin and properties of the bow and arrow. Specifically he is studying why the short Turkish bow was apparently superior to the English long bow in the skirmishes of the Crusades. He has dozens of possibly pertinent books and articles in his memex.

First he runs through an encyclopedia, finds an interesting but sketchy article, leaves it projected. Next, in a history, he finds another pertinent item, and ties the two together. Thus he goes, building a trail of many items. Occasionally he inserts a comment of his own, either linking it into the main trail or joining it by a side trail to a particular item. When it becomes evident that the elastic properties of available materials had a great deal to do with the bow, he branches off on a side trail which takes him through textbooks on elasticity and tables of physical constants. He inserts a page of longhand analysis of his own. Thus he builds a trail of his interest through the maze of materials available to him.   (Bush, 1945)

Douglas Engelbart read the condensed article of 'As We May Think' while he was serving in the US Navy in the Philippines, and it inspired him to pursue a career developing better communication and knowledge tools (Barnes, 1997: 17; Naughton, 2000: 215). In the Navy Engelbart served as an electronics technician, taking care of radios, sound navigation ranging equipment (later known as sonar), teletype transmission, and radio detecting and ranging equipment (radar). After reading Bush's article and returning from the Navy, Engelbart went back to school to finish his degree in electrical engineering. He graduated in 1948, took an engineering job at the Ames Research Laboratory in Mountain View, California, and after a few years got engaged and married. The Monday after his wedding, Engelbart realized that he had achieved all of his life goals—got an education, a steady job, and was married. Now, at the age of 25, he began to think about a new goal for his life (Barnes, 1997: 17). He later described his decision-making process as follows:

I dismissed money as a goal fairly early in the decision process. The way I grew up, if you had enough money to get by, that was okay; I never knew anybody who was rich. But by 1950, it looked to me like the world was changing so fast, and our problems were getting so much bigger, that I decided to look for a goal in life that would have the most payoff to mankind. (Engelbart, quoted in Barnes, 1997: 17)

Engelbart's experience with radar led him to think of interactive computing as a way to implement Bush's memex. Although Engelbart had never worked with a computer, he had read about them in books. He understood that if computers could show information on punched cards and paper, they could also write and draw on display.

I had an image of sitting in front of a display and working with a computer interactively. I had been a radio and radar technician during World War II, so I knew that any signals that came out of a machine could drive any kind of hardware—they could drive whatever you wanted on a display. But I really didn't know how a computer worked. Still, I thought, 'Boy! That's just great!' The images of the different symbologies that you could employ, and other people sitting at workstations connected to the same complex, and working in a close, collaborative way. (Engelbart, quoted in Barnes, 1997: 17)

It took several years before Engelbart was able to implement his vision. First he quit his job and went to the Graduate School of Electrical Engineering at the University of California at Berkeley. At that time, digital computers were just emerging, and his visions of interactive computing didn't get much support. Engelbart decided to change the focus of his research, and earned his Ph.D. in 1955, along with half a dozen patents for bistable gaseous digital devices. After leaving the university, Engelbart accepted a job at the Stanford Research Institute (SRI). During his first two years at SRI Engelbart worked with magnetic computer components, fundamental digital device phenomena, and miniaturization scaling potential. In the process, he earned a dozen patents, and in 1959 he had gained enough of a reputation to begin to pursue his own research interests. With support from SRI and a grant from the US Air Force Office of Scientific Research, he started to work on a paper that described his ideas on interactive computing (Barnes, 1997: 18).

Barnes describes Engelbart's difficulties in setting up research in a new area:

At first, he tried to find an established discipline as a basis for the framework of his design. But people in other disciplines, such as documentation and artificial intelligence, were not interested in his ideas. Finally, he discovered a RAND Corp. report written by J. L. Kennedy and G. H. Putt titled 'Administration of Research in a Research Corporation'. The thesis of the report argued that when a researcher starts an inter- or new-discipline project, the researcher would encounter difficulties when approaching individuals in established disciplines. Each discipline has its own unique conceptual framework that new members of the profession begin to learn during the first year of professional school. If a conceptual framework did not exist for a new-discipline research project, then an appropriate framework must be created. After reading this report, Engelbart started developing his own unique conceptual framework for designing interactive systems. (Barnes, 1997: 18)

In 1962 Engelbart finished his paper. It was titled 'A Conceptual Framework for the Augmentation of Man's Intellect'. A condensed version was published the next year as a chapter in a book, '*Vistas in Information Handling*' (Engelbart, 1963).

Engelbart argued that human cultures had developed systems of spoken and written language, tools and organizing methods which their members needed to solve complex problems in their everyday life. Many of these were learned during individual development, and were specific for each culture. According to Engelbart, those tools and methods that humans used in their problem-solving processes could be characterized as 'augmentation means'. These augmentation means could, in turn, be categorized in four basic classes. *Artefacts* were physical objects designed to provide for human comfort, the manipulation of things or materials, and the manipulation of symbols. *Language* was the way people classified the picture of their world, and provided the symbols that were attached to concepts and used in consciously manipulating the concepts, i.e. in thinking. *Methodology* comprised the methods, procedures, and strategies with which an individual organized goal-centered problem-solving activity. *Training* consisted of the conditioning needed for the individual to bring his or her skills to the point where they were operationally effective for using the three augmentation means (Engelbart, 1963).

In this context, computers could be seen as one among many existing augmentation means. Computers, however, provided qualitatively new opportunities to combine humans and thinking tools. According to Engelbart, computer processes could be developed to match human processes and this would result in a new fourth stage in the development of human intellectual capabilities. The first stage, according to Engelbart, was concept manipulation, the mental capability to use conceptual abstractions of a particular situation and to 'think'. The second stage appeared when these concepts were represented in the human mind as abstract symbols, and these symbols became the objects of thinking. The third stage emerged when humans learned to externalize these symbols, and operate with these externalized symbols, for example, by using graphical representations. The tools used for external representation, however, were mainly based on current ways of manipulating concepts and the present ways of thinking. Engelbart also noted that the tools used in thinking influence the way we think. For example, the language we use constrains the ways we see the world. As a result of these considerations, Engelbart was convinced that new ways to manipulate symbols could lead to new ways to think. To do this, computer processing and human thinking, however, needed to be linked by interactive computing.

Licklider arranged funding for Engelbart in early 1963, and requested that his project be connected to the SDC time-sharing project in Santa Monica, California. Later that year, an online link was set up between SRI's minicomputer in Palo Alto and Santa Monica. Although the funding from ARPA was enough to start the project, Engelbart was still looking for more funding to get the implementation work going. Robert Taylor, who was working at NASA headquarters, had met with Engelbart a couple of months earlier, read his report, and gone out to seek funding for Engelbart. His search proved successful and with ARPA and NASA funding Engelbart was able to start to build a working model of his system (Barnes, 1997: 21).

Engelbart set up the Augmentation Research Center (ARC) at SRI. The primary goal of the Center was to invent technologies that would support individual knowledge workers, and the secondary, collaboration within groups of knowledge workers. Engelbart's team developed a system called NLS (oNLine System), which included features such as email, computer conferencing, graphical user interfaces with multiple windows, hypertext, expanding and collapsing document outlines, videoconferencing, and a mouse. The system was demonstrated in the ACM/IEEE-CS Fall Joint Computer Conference in San Francisco in December 1968. The demonstration used a twenty-foot video screen, two microwave links between the conference location in San Francisco and SRI, thirty miles away, video cameras, and real-time links between the SRI computer running NLS and Engelbart's console on the stage. By comparing these techniques with the prevailing technologies, it is easy to understand that the demonstration created a lot of excitement. The computer industry, however, was not interested in funding Engelbart's work, as his concepts were considered to be impractical. As Barnes notes:

Impressive as Engelbart's demonstration was, the commercial computer establishment did not think his vision of interactive computing was practical. They did not think it was realistic, because the demonstration represented a paradigm shift from batch processing punched cards to direct interaction. Despite the lack of support from the computer establishment, the NLS project continued to develop at ARC with funding from ARPA.   (Barnes, 1997: 24)

4.7 TIME-SHARING AND ON-LINE COMMUNITIES

The vision of interactive computing was closely related to the idea of time-sharing. In practice, interactive use of computers meant that each user needed to use the machine as if it were completely controlled by the user. As computers were extremely expensive, the only way to give users the illusion of having their own computers was to use an operating system that could switch between different users so fast that everyone believed they were connected to the computer all the time. Time-sharing was necessary to efficiently use the expensive computing power, and it was even more necessary when the use was interactive.

In 1961 MIT had started a project to develop a time-sharing system, CTSS. CTSS was an experimental system that could connect four users (Corbató, Merwin-Dagget, and Daley, 1961). Its follower, Project MAC, became the most influential time-sharing project in the 1960s (Lee, Fano, *et al.*, 1992). In 1962 Licklider and Robert Fano, from MIT, travelled together from a computer conference. During the train journey Licklider convinced Fano that MIT's timesharing activities could be expanded and transformed into a study on interactive computing. Over the next weeks Fano prepared a proposal for an ARPA-funded project. According to the proposal:

Computer technology has been progressing by leaps and bounds over the last decade . . . On the other hand, the development of techniques for exploiting computers in non-numerical information processing, and as aids in research and in human problem solving and decision making, has been relatively lagging. Specifically, computer systems . . . have not yet been developed that are *easily* and *economically* accessible, and that are truly flexible and responsive to individual needs, particularly the need for quick, direct response.

An 'on-line' mode of operation in which the individual scientist, problem solver, or decision maker is tightly coupled with a computer system of very large memory and speed appears attractive. It appears even more attractive as we envision the evolution of such a system to provide ready communication with others through machine information retrieval, including the development and use of open data files and public subroutines. On the other hand, in order for any such system to be economically feasible, the machine's memory and processing capacity must be shared simultaneously and independently by many on-line users in such a way as to insure its continuous, efficient exploitation. General-purpose, independent, on-line use of computers by a large number of people has not yet been achieved, but is appears feasible on the basis of recent experiments with time-sharing of large machines.   (Fano, 1963)

The MAC system became available in the autumn of 1963. After several months of experience, its status was described in a progress report (David and Fano, 1965). The report noted that the system is in use 24 hours a day, seven days a week. The total number of user-hours was approximately 17 times the number of computer hours used. The system was usually fully loaded during the day, with 24 on-line users, and it was very seldom idle, even in the early morning hours. According to Fano and David, enthusiasm, however, mixed with a great deal of frustration:

The system was very quickly accepted as a daily working tool, particularly by computer specialists. This quick acceptance, however, was accompanied by the kind of impatience with failures and shortcomings that is characteristic of customers of a public utility. The capacity of the system is limited, and therefore users are often unable to login because the system is already fully loaded. Furthermore, the system may not be in operation because of equipment or programming failures, just at the time that one was planning to use it. In other words, the system is far from being as reliable and dependable as a utility should.   (David and Fano, 1965).

As many of the users of the MAC were computer scientists, they often developed their own solutions to common problems. The users could share files and programs that they had developed for their own needs. Soon such sharing became so common that programmers started to design their programs with the explicit goal of making them easy for others to use. This was a radical departure from the traditional way of programming. Earlier programs had been solutions to unique programming problems. Robert Fano discussed this surprising development in a conference paper a couple of years later:

Some of the most interesting, yet imponderable results of current experimentation with time-sharing systems concern their interaction with the community of users. There is little doubt that this interaction is strong, but its character and the underlying reasons are still poorly understood.

   The most striking evidence is the growing extent to which system users build upon each other's work. Specifically, as mentioned before, more than half of the current system commands in the Compatible Time-Sharing System at MIT were developed by system users rather than by the system programmers responsible for the development and maintenance of the system. Furthermore, as also mentioned before, the mechanism for linking to programs owned by other people is very widely used. This is surprising since the tradition in the computer field is that programs developed by one person are seldom used by anybody else . . . The opposite phenomenon seems to be occurring with time-sharing systems. It is so easy to exchange programs that many people do indeed invest the additional effort required to make their work usable by others.   (Fano, 1967)

Fano further argued that a time-sharing system can quickly become a major community resource, and that its evolution and growth depend on the inherent capabilities of the system as well as on the interests and goals of the members of the community. A system without a display, for example, could discourage development of graphical applications, or the difficulty of several people to interact with the same application could discourage some educational uses. Moreover, Fano noted that after a system starts to develop in a particular direction, work in those directions is preferred and accelerates the development in this direction. As a result, 'the inherent characteristics of a time-sharing system may well have long-lasting effects on the character, composition, and intellectual life of a community'.

4.8 IPTO: TRANSLATING IDEAS INTO MONEY AND TECHNOLOGY

Interactive computing, somewhat unintentionally, changed the way computers were used and programmed. Time-sharing systems opened a new space for collaborative development of technology. In this new domain the benefits of programming were multiplied and an economy of positive returns emerged. Someone, however, had to make the first investment. Institutionally, it was ARPA. Today it may be fair to say that ARPA's Information Processing Techniques Office, IPTO, is historically one of the most successful technology policy initiatives. One of the great mysteries in the history of the Internet is this: how is it possible that one government office played such an important role in transforming the world?

By 1964, when Licklider left ARPA and its Information Processing Techniques Office, the themes of interactive computing and on-line use of time-sharing systems were strongly ingrained in the projects funded by IPTO. When Licklider was preparing to leave IPTO, he suggested that a young computer superstar, Ivan Sutherland, should be nominated to head IPTO. Licklider had met Sutherland for the first time during a conference on interactive graphics, sponsored by IPTO. During the conference, Sutherland asked 'the kind of question that indicated that this unknown young fellow might have something interesting to say to this high-powered assemblage' (Norberg, 1996: 45). Licklider reorganized the next day's program to include a presentation from Sutherland, and the presentation was a great success. When Sutherland became the director of IPTO he was 26 years old. Although Licklider was unsure whether such a young person could measure up to the job, the head of ARPA, Robert Sproull, argued that there should be no problem if Sutherland really was as bright as he was said to be (Norberg, 1996: 45).

Sutherland attended Carnegie Mellon University and Caltech before receiving his Ph.D. in electrical engineering from MIT in 1963. In MIT he worked with Claude Shannon, Wesley Clark, and Marvin Minsky. While working at Lincoln Laboratories, Sutherland developed the first interactive graphics system, Sketchpad.

One of Licklider's principles for funding projects had been that good research results require committed and competent researchers. Sutherland adopted the same approach:

. . . the caliber of people that you want to do research at that level are people who have ideas that you can either back or not, but they are quite difficult to influence. In the research business, the researchers themselves, I think, know what is important. What they will work on is what they think is interesting and important ... Good research comes from the researchers themselves rather than from outside.   (Sutherland, 1989; quoted in Norberg, 1996: 45)

Licklider's emphasis on interactive computing continued when Robert Taylor left NASA to join IPTO in 1965. Taylor and Licklider had met when Taylor was invited to be the NASA representative in an informal committee organized by Licklider. The committee brought together government program officers from the different agencies that supported computer-related research. Taylor had done work with acoustics, as a systems design engineer, and with human-machine systems research, and he had been greatly impressed by Vannevar Bush's article on memex, as well as Licklider's ideas on human–machine symbiosis (Norberg, 1996: 46; Naughton, 2000: 82–5).

When Taylor arrived at IPTO he served eighteen months as a deputy director, sharing work with Sutherland. When Sutherland left and Taylor became the director, he started to develop a program on computer networking. Already in 1965, Taylor had funded an experiment to test the possibility of using long-distance telephone lines to carry bits. One node was at Lincoln Laboratory, on the East Coast, and the other at SDC, in Santa Monica, California. As both places were funded by IPTO, it was easy for Taylor to convince people at both ends that an experiment would be useful. The results showed that cross-continental connections could work in principle, but that noise in telephone lines made it difficult to build practical systems (Naughton, 2000: 86–7).

After becoming the director of IPTO, Taylor was struck by the fact that he had separate terminals to all three laboratories where IPTO supported time-sharing projects. One console was used to connect to the MIT system, another to SDC at Santa Monica, and a third one to University of California at Berkeley:

Three different terminals. I had them because I could go up to any one of them and log in and then be connected to the community of people in each one of these three places ... Once you saw that there were these three different terminals to these three different places the obvious question that would come to anyone's mind: why don't we just have a network such that we have one terminal and we can go anywhere we want.   (Taylor, quoted in Naughton, 2000: 84)

As the IPTO-funded researchers were asking for increasing amounts of computer power for their projects, it was clear that networking could save a lot of money. With a network that could connect all research sites, the machines could be shared by the researchers. If a common computer network could be developed, the different, and incompatible, machines could be made available for all IPTO-funded research projects.

In February 1966 Taylor went to the office of ARPA's director, Charles Herzfeld, and told him about his idea. Later Taylor recalled:

And he liked it right away and after about twenty minutes he took a million dollars out of someone else's budget (I don't know whose to this day) and put it in my budget and said, 'Great. Get started!'   (Quoted in Naughton, 2000: 84)

4.8.1 The Arpanet

Taylor needed a program manager for his network project, and he had an obvious candidate, Lawrence Roberts. Roberts had been working at MIT's Computer Center while a student. When Lincoln Laboratory's TX-O computer, one of the first transistorized computers in the world, was moved to MIT's Computer Center, Roberts started to write programs for it. Soon he worked as a staff associate at Lincoln Laboratory, working with its TX-2 computer. Roberts wrote parts of its operating system and compilers for various programming languages. He also did a Master's degree on data compression and a Ph.D. on computer perception of three-dimensional solids. At some point in time, he wrote a handwriting-recognition program based on neural networks (Norberg, 1996: 46; Naughton, 2000: 85).[9]

As a result of his programming activities, Roberts had become one of the best experts in the TX-2 time-sharing system. When Wesley Clark and Bill Fabian, who had been responsible for the machine, left Lincoln Laboratory after a dispute about whether they were allowed to bring a cat on to the premises, Roberts found himself in charge of the project (Naughton, 2000: 86).

The actual details of Roberts's early interest in computer networks are somewhat unclear. He has said that he first started to be interested in the idea of connecting computers in 1964, when he was invited to join a conference on the future of computing. The conference was held in Homestead, Virginia, and hosted by the Air Force. Many of the ARPA-funded computer researchers were there, and the incompatibility of computers and programs was a topic that came up frequently. According to Roberts:

So, what I concluded was that we had to do something about communications, and that really, the idea of the galactic network that Lick talked about, probably more than anybody, was something that we had to start seriously thinking about. So in a way networking grew out of Lick's talking about that, although Lick himself could not make anything happen because it was too early when he talked about it. But he did convince me it was important.   (Roberts, quoted in Naughton, 2000: 86)

The ARPANET project, in itself, was something that IPTO had not done before. The usual mode of funding in ARPA was to give contracts for research, and not to manage the projects. Taylor, however, wanted to link the different IPTO-funded computer sites, and it was natural that this kind of IPTO-wide project should be coordinated and managed by IPTO. This meant that IPTO was not only deciding how much money to give to its contractors, but it actually had to manage the

---

[9] Naughton notes that Roberts's work on neural networks was 'way ahead of its time'. Naughton probably refers to the boom in neural networks that started in the late 1980s. In the early 1960s neural networks were, however, quite well known. The first neural network models were developed by Nicolas Rahsevky, in the early 1930s, and they became popular after Rashevsky's student Walter Pitts developed a model of logical neural networks in 1943, together with Warren McCulloch. Several computer-based neural network models were developed in the 1950s, and they became a hot topic in 1962, after Frank Rosenblatt published his book on perceptrons. The different models are described, for example, in Olmstedt (1998).

development project. Taylor, therefore, needed a project manager who could make this happen. After trying to hire Roberts for several months, he finally asked help from the Director of ARPA, who called the Director of Lincoln Laboratory. Soon after the phone call Roberts accepted the invitation to come to IPTO to head the network project (Naughton, 2000: 87–8).

Roberts joined ARPA in late 1966. His first plan for the network was based on connecting time-sharing computers at the research sites using dial-up telephone lines. The computers would act as the nodes of the network, transmitting and switching messages from one machine to another. Roberts presented his ideas at a meeting for the principal investigators of the IPTO-funded sites, in April 1967. The site representatives, however, were not excited about the possibility of building a network. One reason was that if the time-sharing computers were to act as network nodes, each node had to run programs that made this possible. It was not obvious why such extra work and use of expensive computer resources would be useful, especially from the point of view of principal investigators who already had computers available. Another problem was that the IPTO plan assumed a complex network of connections between the computers.

The latter problem was solved by Wesley Clark, the principal investigator from Washington University in St Louis. Based on the discussions at the meeting, he realized that the complexity of the networking problem could be reduced by introducing separate interface computers. If every computer had to be able to communicate with every other computer in the network, which were all different, the switching program in hosts would become very complex. If, however, a separate interface machine could be built, a simple network could be designed. New hosts could be connected to the interface machines in a well-defined way, without changing program code in other hosts. This, in practice, meant that the hard work of designing and building the network could be done independently, without much resources from the IPTO-funded laboratories. The managers of the time-sharing host machines simply had to program their machines so that they could talk with the standardized interface processors.

Wesley Clark proposed his idea to Roberts immediately after the conference. Roberts and Taylor accepted it. The interface computer acquired the name Interface Message Processor (IMP) in a report which Roberts wrote and distributed among the Principal Investigators.

This idea of building a separate network connected to host computers through the IMPs was a critical factor in the successful implementation of ARPANET. It enabled IPTO to contract the design and implementation of the network as an independent task, without requiring active collaboration among all host sites. By changing the architecture of the planned system, IPTO was able to set up and fund a new developer community and limit the change in the potential user communities.

In November 1967 Roberts organized a meeting with the IPTO research contractors to discuss the design of the network. This network working group consisted of people from the University of California Santa Barbara, SRI, the University of Utah, and the University of California Los Angeles. Paul Baran was recruited to advise the group. The group's objective was to give criteria and procedures for testing the acceptability of the network and its configuration. Over the next seven months the group circulated a series of working notes and draft specifications. Roberts also awarded a four-month research contract to SRI in December 1967 for a study on the design of the network. In June 1968 Roberts submitted the plan his group had worked out to ARPA director Herzfeld, and the next month he received an initial development budget of $2.2 million for the project. IPTO issued a request for proposals based on the group's specifications in August 1968 (O Neill, 1995: 79; Abbate, 1999: 56).

The request for proposals was sent to potential contractors. The first responses came from IBM and the Control Data Corporation, both of whom declined to bid on the ground that the network could never be cost effective (Naughton, 2000: 131).

The successful bidder was BBN. It had the best proposal partly because it started writing it several months earlier than other bidders. Several key persons at BBN had been working with Roberts at MIT and Lincoln Laboratory and were aware of his networking interests. Robert Kahn, who came to BBN from MIT's Research Laboratory of Electronics where Roberts did his Ph.D., had written a letter to Roberts in early 1967 suggesting that computer networking could be of interest to IPTO (Norberg, 1996: 47). Roberts invited Kahn to a meeting in Washington to discuss the network ideas. Kahn recalled later:

I found out at that point that Roberts was actually interested in creating this net. Having been a mathematician or theoretician, it really had not occurred to me at that point that I might ever get involved in something that could become real!   (Quoted in Norberg 1996: 47)

According to Naughton, based on this knowledge of IPTO's plans, BBN decided to invest $100,000 on preliminary design work in the hope that it would land them the contract:

It turned out to be a terrific bet. By the time the ARPA Request for Proposals eventually arrived in August, BBN was already up to speed on the subject. In thirty frantic days, Heart, Kahn and their colleagues produced a 200-page bid which was by far the most detailed proposal submitted to the Agency. It was, in effect, a blueprint for the Net. The award of the contract was a foregone conclusion. All that remained was to build the thing.   (Naughton, 2000: 133)

Without such a lead time, it might have been more probable that the contract would have been given to a large defence contractor. Indeed, Naughton (2000: 132) notes that the 'general expectation was the contract would go to Raytheon, a major Boston defense contractor, and if ARPA had been a conventional Pentagon agency it probably would have gone that way'. The normal procedure in ARPA was to make contracts directly without competitive bidding. Competitive bidding, combined with the preparatory work at BBN and its tight links with the IPTO design work, made it relatively easy for BBN to get the contract, even though at that time it was a relatively small consulting company whose earlier reputation was mainly in acoustics research. BBN also had close ties to the Honeywell Corporation, whose H-516 minicomputer was a strong candidate for the IMP hardware, and BBN and Honeywell had already decided that they would collaborate in the event that BBN's bid was accepted (Abbate, 1999: 57). The close personal links between

BBN researchers and Roberts also meant that Roberts had a good understanding of the skills available at BBN, as well as the interests of BBN researchers.

The contract was awarded in December 1968, and the work started in January 1969. According to the contract the first IMP was to be delivered to Leonard Kleinrock's laboratory at UCLA in September, and the next nodes at Santa Barbara, Palo Alto, Utah, by the end of the year. BBN kept to the schedule.

When the first four nodes were connected, Roberts planned to extend the network to fifteen computer sites funded by IPTO, and then to additional ARPA research sites, and possibly to military sites. Roberts contracted a newly formed Network Analysis Corporation (NAC) to help with planning. NAC was headed by Howard Frank. Frank was known to Kleinrock through their common lecturing at UC Berkeley, and they shared interests in modelling networks. Frank had been doing work with optimizing the layout of oil pipelines, and had formed NAC to provide services for businesses that wanted to build complex systems. Kleinrock, in turn, had introduced Frank to Roberts (Abbate, 1999: 58).

As the network started to expand, it also became increasingly important to manage the day-to-day operations of the network. According to the BBN contract, BBN was responsible for keeping the IMP network running. The BBN team soon found out that the operation of the network was a major effort in itself. When BBN's own node became connected to the network in 1970, IPTO funded BBN to set up the Network Control Center (NCC) to manage the operation of the network (Abbate, 1999: 64–6; O Neill, 1995: 79).

The growing network also meant that there was an increasing need to share information about developments in the project. For this purpose IPTO awarded a contract to SRI to set up the Network Information Center (NIC). SRI was selected partly because Engelbart's group was located there, developing systems for information sharing.

5

Analysis of the Early Phase of Internet Development

When the year 1969 turned to a new decade, there were four computer nodes in a network that later became known as the Internet. Just three months earlier the first two nodes of the ARPANET connected for the first time. Could anyone, at that time, have known whether this network would still exist at the end of the new decade? Or would it just remain one of the unfulfilled dreams of computer networking?

As the previous chapter showed, there were four loci of concentrated innovation, which provided the platform for growth. The first of these was formed around the idea of interactive computing and the augmentation of human intellect. Its origin was in a theoretical view of humans as information processors and self-regulating cybernetic systems, combined with new technologies of radar displays, storages that could record massive amounts of documents, and time-sharing digital computers. Vannevar Bush gave this movement legitimacy and inspiration, J. C. R. Licklider funded it, and Douglas Engelbart gave it a material form. The vision was to build machines that could help individuals and communities to think and make decisions in an increasingly complex world.

The second locus of innovation was located at RAND. Paul Baran and his colleagues did not want to increase human intelligence. They had a more modest goal of trying to avoid big mistakes that could happen if military command and control were destroyed. Whereas Bush, Licklider, and Engelbart were interested in thinking, Baran's keyword was communication.

The third locus of innovation was at the National Physical Laboratory in the UK. Donald Davies was envisioning a network that could be used for business and pleasure. He believed that time-sharing computers would make ordinary citizens computer users, and that in the process new services and industries would be created.

The fourth locus was the ARPANET project. Its underlying vision was Licklider's Galactic Computer Network and a belief in augmenting human capabilities with computers. The ARPANET vision also benefited from experiences in time-sharing computing, which indicated that programmers could form collaborative communities when a computer was used to share information and programs. Its key rationale, however, was the belief that networking time-shared computers could lead to great savings in resource use.

With the benefit of hindsight these four loci of innovation could be character-ized in the following way. Engelbart's system was a tool for knowledge workers. Baran developed distributed communication. Davies's system was infrastruc-ture for network society. ARPANET was trying to develop a system for distributed computation.

5.1 TECHNOLOGICAL FRAMES

To understand innovation in computer networking, it is useful to make a short digres-sion to the history of plastics. Based on his studies of the history of technology, including plastics, Bijker argued that the evolution of technology can be described as the evolution of *technological frames*. A technological frame, according to Bijker (1987; 1997), is not something that is located in an individual's mind, institutions, or technological systems. Instead, technological frames apply to the interaction of vari-ous actors. A given technological frame is linked to a technology and the way its developers understand it. But the locus of technology development is not simply in the developer community: it is also linked to user communities and their practices.

According to Bijker, technological frames include current theories, goals, problem-solving strategies, key problems, tacit knowledge, design methods, testing proced-ures, exemplary artifacts, and practices of use (Bijker, 1987: 171; 1997: 122–7). A given frame brings with it a system of practitioners and practices that provide the context where a new technology is interpreted and applied. Bijker, for example, showed that much of the early work in plastics in the nineteenth century occurred in the frames of the dye chemistry and by entrepreneurs who were looking for substitutes for ivory. The success of Leo Baekeland in developing an industry around Bakelite depended critically on his ability to work between separate technological frames, in effect redefining what plastics were about. Furthermore, his success also depended on his skill in recruiting key members of the competing frames into his own new frame, and on his skill in rapidly expanding and redefining it by linking it with such newly emerging industries as automobile and radio manufacturing.

Bijker argued that Baekeland's success was possible because people can have different degrees of *inclusion* in technological frames. Technology is interpreted in different ways through different frames, and those who are at the centre of a frame are relatively tightly bound by a network of practices, theories, goals, and interpreta-tions. Although Baekeland had a relatively low inclusion in the technological frame of plastics, he had a high inclusion in the frame of electrochemical engineering. When his early attempts to solve problems of plastic-making failed, he switched to a problem-solving strategy that was used in the electrochemical frame. Many other inventors were trying to solve the problems of plastic manufacturing. They were, however, looking for the solution from within the existing frame of plastic technology, and failed.

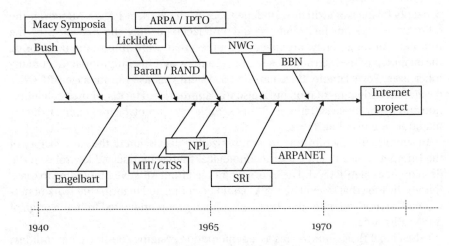

Fig. 5.1. Some key loci of innovation in the evolution of the Internet, 1940–1975

As we saw above, computer networking had several underlying technological frames, which had become relatively stable around the mid-1960s. At that point, computer networks were imaginary. They did not exist as material artefacts or systems. The 'product' was mental and its users were its developers.

Packet-switching networks were first implemented at the end of the 1960s, and the first internetworks were designed a couple of years later. By looking backwards from that point in time, we can ask: how did we get there? Where do we find the loci of innovation that made the history of the Internet? Even a simplified picture looks quite complex, as can be seen from Fig. 5.1.

As we are used to the linear model of innovation, Fig. 5.1 is easily read as a model of causal links. Indeed, it is interesting to see how difficult it is to find a pictorial representation that locates events in time and does not suggest causality. When we position events in time, it becomes almost impossible to do it in a way that does not imply that 'events follow one another' and that one event 'leads to another'. Fig. 5.1, therefore, looks 'wrong'. It suggests that, for example, Paul Baran's work at RAND 'led' to the emergence of the Internet. The correct interpretation, of course, is that in the evolution of the Internet, Baran's work and the resources it created were retrospectively appropriated at several points of time, including the early phase of the ARPANET design. The linear model is deeply ingrained into our way of thinking—and even the way we draw pictures—and therefore it looks natural.[1]

The different loci of innovation, and the resources they created, however, did not determine the unfolding of events. Instead, they facilitated development. If Baran's work had not been available, all evidence indicates that other resources could have been used. For example, Davies's work at NPL would have provided the basic

---

[1] As Eliade (1991) pointed out, modern Western culture organizes its event historically. This historicity implies causality and progress.

concepts for packet-switching, independent of the work at RAND. Moreover, the different innovation loci depicted did not facilitate only the evolution of the Internet. Although they produced resources that were important in shaping the evolution of the Internet, these same resources were also appropriated for many other uses. For example, Engelbart's NLS system was used to manage ARPANET development documentation but it also was appropriated for distributed collaboration and hypermedia systems, and, for example, as a prototype for graphical direct-manipulation user interfaces.

To simplify this picture, and to summarize the discussion of the early history of the Internet, we can define four technological frames that played key roles in the development of ARPANET. The Internet emerged from ARPANET by adding to these frames the idea that several networks could be connected to form a network of networks. The motivation for this, however, in many ways originated from the original ARPANET frames.

I shall label these four frames as 'augmentation systems', 'on-line communities', 'communication networks', and 'electronic services'.

5.1.1 The Augmentation Systems Frame

Historically, the oldest frame is the augmentation systems frame. Its key idea was that human capabilities could be enhanced by tools, and that such tools could also support mental work. This frame was conceptually developed in a series of Macy Foundation conferences, and by visionary scientists, in the 1930s and 1940s. As was noted before, its key assumptions were recorded in Vannevar Bush's article 'As We May Think', published in 1945, Norbert Wiener's *Cybernetics*, published in 1948, in J. C. R. Licklider's 'Man-Computer Symbiosis', published in 1960, and in Douglas Engelbart's 'A Conceptual Framework for the Augmentation of Man's Intellect', published in 1963.[2] In 1965 its locus of activity was in SRI's Augmentation Research Center, and its reference implementation was the NLS system. This frame is depicted in Table 5.1.

The augmentation systems frame was a combinatorial frame. A previous practice that most closely implemented its ideas was the SAGE air-defence system, which connected time-sharing computers, radars, telephone and radio links, and humans into an early-warning system. Radar was also an important predecessor, as it showed how graphical displays could be used for interactive and cognitive use of technology. In the early phase of the development of this frame, microfilm technology was an important reference technology, as it showed how large amounts of documents could be stored effectively, possibly in a form that could be mechanically manipulated.

---

[2] In the US, the origins of this frame could also be found in the cognitive theories of pragmatists, including James, Dewey, and Mead.

Table 5.1. *The Augmentation Frame, 1965*

Frame attributes	Frame content
Theory	Cybernetics
Techniques	Interactive modelling, interactive graphics, information storage and retrieval
Use	Information and knowledge management, decision support, collaboration support
Thought leaders	Vannevar Bush, Ross Ashby, Jay Forrester, Oliver Selfridge, Douglas Engelbart, J. C. R. Licklider
Sub-frames	Microfilm, radar, neural information processing, SAGE
Locus of activity, 1965	SRI Augmentation Research Center

Table 5.2. *The on-line communities frame,* 1965

Frame attributes	Frame content
Theory	—
Techniques	Time-sharing operating systems, on-line terminals, file sharing, email within a single time-sharing computer
Use	Effective use of large and expensive computers, 'synergy'
Thought leaders	J. C. R. Licklider, Douglas Engelbart, Fernando Corbató, Robert Fano, Robert Taylor
Sub-frames	SAGE, Augmentation systems
Locus of activity, 1965	MIT, SRI, ARPA/IPTO

5.1.2 The On-line Communities Frame

The second frame was the on-line communities frame. It grew out of the motive to efficiently utilize expensive computer resources. It therefore became possible only after computers existed and were regularly used. Early on, it was noted that when several users were connected to the same computer, the users formed a community where work and its results could be shared. Although it was at first expected that the key benefit of building time-sharing operating systems and on-line terminals was more efficient use of computers, it soon appeared that such a community of users became more than the sum of its computer resources and individual members. The on-line communities frame is depicted in Table 5.2.

Several observations can be made about Table 5.2. First, this frame did not have any clear theoretical tradition. Instead of a well-developed theory, it was based on a common-sense economic conception that effective use of an expensive and critical resource would be useful. The belief in synergy was at least partially based on the experiences of the system designers: the use of the system 'surprised' them, and therefore was 'more' than what was originally intended and designed. For example, as was pointed out above, much of the operating system programs for the early time-sharing systems were developed by their users, independent of the plans of the system mangers. This potential for positive surprises was formulated as the idea that by putting large enough number of users together, synergistic benefits would probably be materialized.

Although the point of origin for the on-line community frame was in the time-shared operating system development community, it also had close links to the augmentation systems frame. Indeed, the first real on-line community system was built by Engelbart's team at SRI's Augmentation Research Center. One could view the augmentation systems framework as linked and partially embedded into the on-line community frame. The on-line community frame, in effect, had two sides: the 'hard' side focused on distributed computing and efficient use of technology; the 'soft' side focused on human collaboration and effective use of skills, knowledge, and work. This linking occurred to an important extent through ARPA's Information Processing Technologies Office (IPTO), which was both funding equipment purchases and development, but also looking for innovative uses of new technology.

As was noted before, IPTO was formed by J. C. R. Licklider after he joined ARPA in 1962. Licklider effectively implemented his vision of interactive computing by funding both Engelbart's work on augmentation systems, and most of the research on time-sharing systems in the US. In effect, Licklider 'modularized' his vision of interactive computing and outsourced its modules to people who were interested in implementing parts of it. Licklider, therefore, emerges as one of the main 'authors' of modern computing and the Internet. His impact was increased and made possible by the fact that several of the subsequent directors of the IPTO were socialized to Licklider's vision. IPTO therefore continued to have a major influence on the development of this vision for over a decade (O'Neill, 1995; Norberg, 1996).

Licklider, himself, combined several technological frames in his personal vision. He was involved with the cybernetics tradition, psychology of perception, mathematical modelling of acoustic phenomena, and also had experience in computer programming. The interest in the time-sharing use of computers was to a large extent generated by the perceived need for interactive computer graphics programming, and this, in turn, was seen as a way to implement the augmentation systems frame. A more technology-oriented time-sharing frame, however, formed early on around the developers and users of time-sharing operating systems. There were several members in this community who did not explicitly subscribe to the augmentation systems frame. Much of the work around the MIT's MAC project, for example, was focused on developing systems that tried to automate cognitive processes, instead of supporting them.

The time-sharing frame, in turn, expanded early on to the on-line communities frame. The use of time-sharing systems showed that the practice was not only about technology, but also about human users. In a sense, there was a spin-off from the augmentation frame to a time-sharing frame, which then rapidly expanded to the on-line community frame.

5.1.3 The Communication Networks Frame

The third technological frame was about communication networks. It originated in military command and control practice, with the goal of developing communication

Table 5.3. *The communication networks frame, 1965*

Frame attributes	Frame content
Theory	Queue theory, traffic modelling, information theory
Techniques	Digitalization, message switching, radio technology
Use	Reliable communication
Thought leaders	Paul Baran, Leonard Kleinrock
Sub-frames	Telephony, telegraphy, telemetry
Locus of activity, 1965	RAND, MIT

systems that could survive nuclear attacks. It was also developed in parallel with the information theory framework. The first line developed architectures and designs for reliable digital computer-based communication networks, and the latter focused on mathematical modelling of information flow in digital networks. These two lines of development were relatively independent until around 1967.

The communication networks frame can be viewed as a combinatory frame, although important parts of it were spin-offs from existing traditions. Paul Baran, working for RAND, developed packet-switching computer networks as a solution to a problem that was easy to understand and much discussed at the beginning of the 1960s. Baran was designing a communication system that could work even after a nuclear attack destroyed much of the communication network (Baran, 1964: vols. I, XI). Baran's problem, however, is a clear example of a *presumptive anomaly* (Constant, 1987). Such a presumptive anomaly emerges when theory predicts that a technological system may fail under future conditions. At the time he was developing his system, communication networks worked fine. There were no atom bombs falling from the sky.[3] Baran was unable to stabilize his frame, as many telecommunication experts rejected the assumptions that were at the core of Baran's design. Baran's work was picked up later, when it was rediscovered by the NPL researchers in the U.K. who were trying to implement computer networks. Soon after that, in 1967, Baran's frame was embedded in the on-line community frame. Similarly, theoretical modeling work that had been done at MIT was embedded in the on-line community frame, which at that time became the core frame for ARPANET developers.

5.1.4 Electronic Services Frame

The fourth frame is the electronic services frame. It emerged in a few months, in 1965–6, when Paul Davies, working at the National Physical Laboratory in the UK,

---

[3] Indeed, Bijker's (1987) analysis of presumptive anomalies would indicate that technological frames are difficult to stabilize around the threat of nuclear bombs. It is interesting to note that for some reason this non-existent nuclear war has often successfully been used to form communities of technological practice. Unrealized opportunities and unrealized threats seem to lead to different types of presumptive anomalies, especially in the case of the threat of war, which probably in many ways is built into social structures, value systems, power relations, and linguistic metaphors.

Table 5.4. *The electronic services frame, 1965*

Frame attributes	Frame content
Theory	—
Techniques	Time-sharing operating systems, remote terminals, digitalization, message-switching
Use	Electronic access to services and entertainment
Thought leaders	Paul Davies
Sub-frames	Telephony, telegraphy, telemetry, time-sharing systems, technology policy
Locus of activity, 1965	National Physical Laboratory, UK

realized that time-sharing computing could provide a platform for a new economy and society. Davies first started from the problem of developing a network that could be used to connect remote users to a time-sharing computer, but he quickly noted that such a network could enable many novel services. In Davies's vision, computer networks could be used as an infrastructure that would provide services for large populations of big cities, regions, for the whole country, and which could eventually develop into a global network. To implement this vision, Davies designed a digital packet-switching network that was in many ways similar to the one developed by Paul Baran at RAND.

An interesting characteristic of the electronic services frame was that it was linked to the frame of technology policy. Davies was interested in developing the UK computer industry in the new promising area of computer networking. His frame therefore naturally evolved into social and economic directions, and the technical design was only a way to implement the frame in practice. Due to the fact that the communication networks were the monopoly of the General Post Office in the UK, Davies ended up in a similar situation to Baran. A new economy and network society were presumptive anomalies from the point of view of Post Office managers. Indeed, we could call Davies's vision a *presumptive opportunity*. As the system envisioned by Davies did not provide a solution to any obvious existing problem, it was difficult to mobilize resources for its implementation. Davies therefore early on redefined his goals, and eventually implemented a system that closely resembles modern local area networks. In other words, Davies started from the electronic services frame but eventually found himself in the on-line community frame.

The interactions between the early frames were complex, and the complexity of these linkages increased rapidly after 1965. The electronic services frame did not stabilize in the 1960s. It was picked up a couple of decades later, largely because technological infrastructure and practice made it relevant again. The communications network frame became embedded both in the ARPANET community and, in the 1970s, in the telecommunications community. The on-line community frame became a central frame for the developers of ARPANET and other research-oriented computer networks, such as Merit, Cyclades, and, later USENET, FIDONET, and

BITNET. In the early 1990s the Internet and World Wide Web integrated some aspects of all these frames, and provided a new platform that accelerated the pace of development further.

5.2 RESOURCE MOBILITY IN THE EARLY PHASES OF INTERNET HISTORY

For an innovator, needs, opportunities, and problems mutually construct each other, and provide the context where the innovator's activity makes sense. The problems that have to be worked on follow naturally from the way opportunities and needs align. For Baran, building the net meant finding ways to route the traffic so that it could get from one point to another securely, even under the most severe conditions. For Engelbart the net meant collaboration among knowledge workers, and the problem of developing tools for the various tasks that this work required. For Davies the problem was simply to get the system and its services working so that it could be rolled out to the rest of the nation and across the globe. For ARPA the problem was to get the ARPANET and its Interface Message Processors implemented, and to get the host computers to communicate with each other.

The context of innovation helps us understand why innovation is happening, and what makes its actors move. To actually make innovation happen, resources are also needed, however. Raw materials, tools, and work need to be organized to give a shape to the emerging technology that embeds the innovation.

It is useful to divide resources into two different kinds. Some resources are 'sticky'. Their accumulation often takes a long time, and they are not easy to reconfigure. Other resources, in contrast, are 'mobile'. They can easily be moved to where they are needed.

Human competence is often surprisingly sticky. It is relatively easy to move people from one place to another, but competences don't move so easily. Competence is very much linked to a configuration of tools, layouts of buildings, information networks, organizational decision-making systems, reputation, trust, social networks, and the way we have learned to construct the world as a set of problems and share it with our colleagues. In other words, competence is often grounded in communities of practice that produce and reproduce competences.

The foundation of competences, therefore, evolves according to the dynamic of social learning. Competence is also often appropriated outside the community where it is embedded. Combinatorial capabilities can be built at the interstices of existing communities. A member of a community can stand for the competences of the community, and competences can be configured in temporary groups.

When we take the underlying communities for granted, it may seem that human competence is highly mobile and that we can easily form temporary groups with any required sets of competences. Mobility of resources, however, often occurs in a context where much of the underlying social interaction has become invisible and institutionalized.

One important form of institutionalization is what we could call social and technical infrastructure. When we accumulate infrastructure for competences, human skill becomes increasingly mobile. For example, if two research laboratories have similar operating procedures, and use similar tools and conceptualization of the world, a researcher can move relatively easily from one laboratory to another without becoming incompetent. By institutionalizing and standardizing some aspects of competence, we can move some components of competence more easily. Technology, itself, can be understood as a way to sediment aspects of competent human activity into material artefacts.

From this perspective we can ask what resources were used in the different loci of innovation discussed above, and what competences were moved into new configurations to overcome perceived problems.

In the case of RAND, it seems that not much moved. Baran's distributed message block networks emerged to a large extent from his own initiative. Baran used his earlier experiences in telemetry and translated his interests so that they fitted with the legitimate goals of RAND. He learned about the opportunity to build survivable networks from earlier store-and-forward networks, and used slack resources and time at airports to sketch a solution to the problem he perceived. But, although Baran didn't move a lot of resources, his work made sense only as a part in a much larger picture. In this picture many pieces moved. Imaginary transcontinental missiles fired across the sky, satellites were shot toward stationary orbits, and somewhere machines loaded nuclear bombs into heavy airplanes and submarines. Without moving much, Baran moved to the centre of this picture.

In the case of Engelbart, resources started to move at the beginning of the 1960s. The infrastructure was provided by SRI, and on top of this infrastructure Engelbart and his team started to build new technology. The funding came from government sources, mainly from ARPA. The system that Engelbart's Augmentation Research Center produced was interesting in itself, and attracted resources, including competent people, for over a decade. Engelbart borrowed and adapted technologies that had been developed for other purposes, making technology do something its original developers didn't intend it to do. Using any criteria for innovativeness, Engelbart's team was extremely innovative. Old technology was modified and adapted to new uses, missing technologies were designed, implemented, and combined with existing systems, and the basic purpose of the system under development was continuously redefined. The problem which Engelbart attacked was so generic that almost anything could be part of the solution. Every improvement was an improvement, and all available resources moved in a continuous recombination and reconfiguration. This, however, was possible only because the environment for this innovative activity was stable. SRI provided physical and organizational infrastructure, and IPTO moved money so that the project stayed alive.

The NPL provided similar infrastructure for Davies's team in the UK. His own position as the superintendent of the computer division gave Davies both legitimacy and opportunity to reallocate resources to get the project started. This beneficial position, however, was also probably the reason why the NPL network never extended outside its place of origin. Davies was able to move resources, but only those resources that were under his own control. The NPL team had many years of experience and competence. The main problem was to put together those pieces of technology that were to provide the services for the network. Due to budget constraints, the resources were limited and development was slow. Within the limits, the NPL team could, however, make progress and feel that the problem was being addressed. At the same time it became very difficult to move the results out of NPL. In effect, the NPL resources became detached from those resource allocation mechanisms that would have been required to make Davies's original vision reality in the outside world. To make that happen, a new locus of innovation should have emerged outside the NPL, moving the ideas and results of NPL work to wider use.

In a sense, this was exactly what happened. The ideas developed at NPL had a clear impact both on the early phases of ARPANET and on the design of its Interface Message Processors. NPL also played an important role in the development of European computer networks and made major contributions in creating packet-switching telecommunications infrastructure.

The ARPANET project, however, became a locus of innovation where technical visions were combined with a unique pool of resources. IPTO had already funded host computers that were in use in what were to be the ARPANET sites. The computers were expensive and heavy, and became core artefacts around which new competences and communities could form. After this infrastructure was built, IPTO was able to move competent people in these sites to think about the problems of the network. In this process the problems became understood from a collective and inter-organizational perspective. As a result, the network was constituted as a network of sites, users, and uses from the beginning. IPTO used its funds and knowledge of available research competences to outsource pieces of the complete puzzle. This effective division of labour made fast progress possible, at the same time building a social network that became the channel for the diffusion of the emerging technology.

The ARPANET program differed from the other network projects in that it had much more money and resources available than the other projects. It did, however, have another crucial difference as well. The ARPANET program very early on evolved into a network of innovation loci. It relied on a distributed innovation process, loosely connected by sharing technological artefacts, documents, and overlapping communities. This distributed architecture became one of its key characteristics already before the first nodes were connected to the net, and it became even more pronounced as new nodes started to be connected. Indeed this multi-focal process of innovation seems to be the key to the success of ARPANET and the Internet.

At the beginning of the 1970s, there was, however, no guarantee that ARPANET would became a successful project. A successful implementation of a technical system rarely means that the system will be taken into use. Moreover, ARPANET was in many ways an unsuccessful system in 1970. Although it did move digital packets from one computer to another with some reliability, there was not much that the network could do. From the point of view of research sites, it was probably more a problem than a solution. If ARPANET had been switched off, not many people would have protested. In fact, very few people would have noticed.

The multifocal innovation process became possible because ARPANET was fundamentally a collective project. Although it was strongly led by Larry Roberts at IPTO, Roberts never became the key innovator in the process. Instead, the program developed new ways of working that quickly became the key factors in the survival of this network project. Indeed ARPANET implemented a distributed survivable innovation network. In this sense it exceeded Paul Baran's original dream. It did not only develop into a robust technical platform that was able to relay messages; instead, it became a distributed socio-technical system where meaning was both created and shared. In other words, it became both a platform and the object of innovation.

This new mode of innovation started to acquire a definite form in the early 1970s. One of the most fundamental loci of innovation formed around an informal group of people. This group became known as the Network Working Group. During the years, it evolved into a community that coordinated the evolution of the Internet, and which provided most of the expertise that made the evolution of the network possible. In many ways, the early experimentation with the processes of the Network Working Group defined a culture of continuous innovation, which enabled the Internet to change its nature several times during the next three decades. I will discuss the characteristics of this group below, and show how it led to a new innovation model.

First, however, it is useful to make another visit to theory and study in more detail the different proposals on conceptualizing the loci of innovation. What, exactly, are these 'communities' and 'loci of innovation', which seem to acquire such a prominent role in the multifocal model of innovation? Several different theoretical traditions become relevant here. The next chapter introduces these theoretical traditions.

6

Socio-Cognitive Spaces of Innovation and Meaning Creation

Innovation has traditionally been understood as something that originates in the acts of an inventor. As has been noted several times in the previous chapters, this rather innocent-looking assumption leads to the linear model of innovation. More refined versions of this model include feedback mechanisms and improve the basic linear model. During the last couple of decades, it has, however, become increasingly well understood that innovation needs to be studied as a social phenomenon, and that innovations are articulated in a social process. The 'locus of innovation', therefore, is something more than an individual inventor. Innovation does not happen 'inside the head' of an inventor; instead, innovations emerge in a field of social interaction.

This chapter explores several different proposals that have tried to characterize such 'fields' and 'spaces' of innovation. Nonaka and Konno (1998) proposed that the locus of innovation could be described as '*ba*', a Japanese concept that can be translated as 'space'. Constant (1987) and Brown and Duguid (2001) argued that the locus of innovation should be viewed as a *community of practice*. Based on cultural-historical activity theory, Engeström (1987; 1999), in turn, argued that the locus of innovation can be found in *activity systems*. Other proposals have included, for example, techno-economic paradigms and trajectories (Nelson and Winter, 1977; Freeman, Clark, and Soete, 1982; Perez, 1985; Dosi, 1982; Dosi, Freeman *et al.*, 1988), technological communities (Van de Ven, 1993), communities of creation (Sawhney and Prandelli, 2000), networks (Powell, Koput, and Smith-Doerr, 1996), techno-economic networks and actor-networks (Callon, 1992), epistemic cultures (Knorr Cetina, 1999), design spaces (Stankiewicz, 2000), regions (Saxenian, 1994), and social practice (de Certeau, 1988). These proposals are complementary, but they all imply that innovation cannot be understood simply as an individual or organizational phenomenon.

In this chapter I will focus on socio-cognitive spaces of innovation and meaning creation. These 'spaces' are the interpersonal fields where new knowledge is generated and articulated as new designs and technological artefacts. I will introduce several different conceptualizations of innovation spaces, focusing on the concepts of thought community (Fleck, 1979), chronotope (Bakhtin, 1987), practice-related communities (Constant, 1987; Lave and Wenger, 1991; Wenger, 1998; Brown and Duguid, 1991; Brown and Duguid, 2001; Schön, 1983), activity systems (Engeström, 1987; Leontev, 1978), and the concept of *ba* (Nonaka and Konno, 1998; Nonaka, Toyama, and Konno, 2000).

A sufficiently detailed understanding of the nature of innovation spaces is necessary if we want to discuss social processes that generate new social practices and innovations. The difficulty in such a discussion is that innovation spaces are not conceptually trivial. Social interaction is complex, reflexive, and highly dynamic. The easiest way to approach the question of defining the nature of innovation spaces, therefore, is to highlight them from different angles, in the process outlining their shape. Although such a description cannot be complete, and does not give a fully integrated picture of the phenomenon, it allows us to point out some important aspects of innovation processes.

One key characteristic of innovation spaces is that they are embedded in social life. In recent knowledge-creation literature, the social foundations of knowing have been discussed extensively. This discussion has to a large extent centered around the concept of community. The concept of community is useful as it naturally leads to the analysis of interpersonal dimensions of knowing, and links knowledge creation with the theory of communication and social learning. The concept of community, however, has its problems. It is loaded with meanings that may become misleading when we discuss innovation. The concept looks familiar and clear, but its use leads us to the very foundations of social and epistemological theory.

For example, Brown and Duguid introduce the concepts of *community of practice* and *network of practice* to discuss innovation in organizations (Brown and Duguid, 2001; Brown and Duguid, 2000: 141–3). They argue that communities of practice are tight-knit groups formed by people working together on the same or similar tasks. According to Brown and Duguid, these should be distinguished from networks of practice that consist of people working on similar practices, but where the members of the network do not necessarily know each other.

In Brown and Duguid's terminology, *networks of practice* are loosely knit 'special interest groups'. Their members do not interact with one another directly to any significant degree, don't take action, and produce little knowledge. Brown and Duguid maintain that the concept of *community of practice* focuses on subsections of these networks of practice. Brown and Duguid note that communities of practice are usually face-to-face communities that continually negotiate with, communicate with, and coordinate with each other directly in the course of work. Moreover, they maintain that the members of communities of practice usually know each other and work together directly. In this process the community develops its own language, thought style, and judgement.

For Brown and Duguid, an important difference between a community of practice and a network of practice is that the former is a locus of action. The examples provided by Brown and Duguid, however, consist of different types of loci for collective action. Their first example is a small product development team consisting of less than half a dozen people (Brown and Duguid, 2000: 127). Indeed, their example is the BBN team that developed the network switching computer for ARPANET.

This group consisted of experts in software design, hardware design, and communication networks. It was a team that combined different expertises in a temporary group that had the goal of creating a common technological artifact. Brown and Duguid, however, also associate the concept of communities of practice to communities studied by Lave and Wenger (1991). Lave and Wenger were interested in social learning processes, and their focus was anthropological and ethnographic. Their examples included communities such as Alcoholics Anonymous groups, Yucatan midwives, Vai tailors, and meat cutters. Whereas the composition of the ARPANET development team was obviously based on division of labour, such division of labour or specialization of knowledge is more difficult to find in an AA group, for example.

The essential characteristic of Lave and Wenger's communities was that they shared the same stock of knowledge, although to different degrees. This shared knowledge was grounded in their shared practice. One could call such communities 'homogeneous' communities (Tuomi, 1999: 271–5). The differences in their memberships are differences in the centrality of their members. The members of such communities construct their identities based on the membership and socialize to the world-view of the community, but there is no clear coordination of work, or collective production. In this sense, the networks of communities, as defined by Brown and Duguid, seem to be very similar to Lave and Wenger's communities of practice.

This ambiguity in terminology is partly caused by an alternating focus between two perspectives: learning and exploitation of knowledge. When the focus is on social learning of an existing practice, community of practice is exemplified by the groups discussed by Lave and Wenger. When the focus is on active co-production, the locus of activity is often found in a subsection of networks of practice, for example, in teams that bring together complementary resources.

More generally, the existing literature seems to have four different concerns. Lave and Wenger were clearly focusing on *communities of identity and competence development*. Brown and Duguid, in turn, use the term community of practice to describe *communities of production*, in a way that closely resembles Constant's description of communities of practitioners (Constant, 1980; 1984; 1987). The Bakhtinian concept of genre, which is described in more detail below, in turn, defines a *community of shared meaning*, which according to Bakhtin is closely linked to social practice (Bakhtin, 1987; Morson and Emerson, 1990). Furthermore, several consulting companies have used the concept of community of practice to refer to a group of people who exploit a specific technology or disciplinary knowledge in their business practice. Such communities might be called *communities of appropriation*.

All these views are important, and impossible to separate in actual social life. To characterize, and hopefully clarify, the nature of communities and innovation spaces we may, however, start from earlier descriptions that have elaborated the above themes. I will first introduce Fleck's work on thought collectives. It is an insightful description of those socio-cognitive processes that underlie the emergence of new knowledge, technology, and practice.

6.1 THOUGHT COLLECTIVES

As a result of his studies on immunology, bacteriology, and experimental medicine, Ludwik Fleck came up with ideas that closely connect development of knowledge with development of practice, social structure, and technology. In 1934 Fleck wrote a historical study on the emergence of syphilis, as a well-defined disease, and argued that scientific knowledge is bound to 'thought collectives'.

According to Fleck, cognition cannot be construed only as a dual relationship between the knowing subject and the object to be known. The existing fund of knowledge provides the basis for the emergence of all new knowledge. What is already known influences the way the world is interpreted. Fleck noted:

Cognition is therefore not an individual process of any theoretical 'particular consciousness'. Rather it is the result of a social activity, since the existing stock of knowledge exceeds the range available to any one individual.   (Fleck, 1979: 38)

Fleck's studies showed how syphilis emerged as a well-defined disease entity, from a relatively undifferentiated mass of symptoms. During this process, which started at the end of the fifteenth century, the nomenclature and differentiation of what we now know as syphilis went through many changes. The conceptualization of syphilis as a disease, various diagnostic practices and tools, scientific descriptions of its nature, and the community of medical practitioners co-evolved during the centuries. Finally, syphilis became associated with the Wasserman reaction.

In the first documented descriptions of carnal scourge, astrology played an important role, as well as ethical arguments that the condition was a punishment for sinful lust. When it was noted that mercury can cure some forms of carnal scourge, syphilis became the specific type of carnal scourge that can be cured by mercury. The therapeutic power of mercury, as well as its diagnostic power, however, remained unsatisfactory until the end of the nineteenth century. The idea of carnal scourge meant that the indications of syphilis included what we today know as other venereal diseases, including gonorrhoea, soft chancre, and their complications. It was therefore argued that sometimes mercury did cure the carnal scourge, while sometimes it just made things worse.

As late as 1890, Dr Josef Hermann, who was for many years the head of the department of syphilis at the Imperial and Royal Hospital of Wieden in Vienna, argued that syphilis does not exist. Hermann maintained that many serious symptoms that were normally considered to be indications of syphilis were 'exclusively produced either by the mercury treatment itself or by other bad concoctions' (quoted in Fleck, 1979: 6). This rejection of syphilis as a specific disease was to a large extent a response to the very diffuse nature of the symptoms. For a long time there was hardly any disease or symptom that was not attributed to syphilis (1979: 12). Given the existing tools and practices for diagnosis, it was almost as easy to argue that syphilis is everywhere as it was to argue that it was nowhere.

Since the first discussions on syphilis, syphilitic blood was offered as a cause of the disease, first as a bad mixture of humors, then as foul blood, as excessively hot and thick blood, and finally as blood befouled by an infection. Eventually, this idea of syphilis as something that can be defined as a change in blood resulted in using the Wasserman reaction as a method to separate syphilis from other conditions. The Wasserman reaction quickly became a topic of much experimentation and academic discourse and created the discipline of serology. In this process, the complex and theoretically not well-known procedures of the Wasserman test became standardized through new concepts, tools, and practices, and social institutions emerged that became the legitimate guardians of these standards and practices.

Fleck noted that the history of syphilis and the Wasserman reaction shows that many different views compete in the evolution of ideas and practices. Scientific facts make sense only within a specific style of thought. There are several styles of thought, and each of these is socially grounded:

If we define 'thought collective' as a community of persons mutually exchanging ideas or maintaining intellectual interaction, we will find by implication that it also provides the special 'carrier' for the historical development of any field of thought, as well as for the given stock of knowledge and level of culture. (Fleck, 1979: 39)

Fleck argued that, although a thought collective consists of individuals, it is not simply an aggregate of them. Instead, the 'individual within the collective is never, or hardly ever, conscious of the prevailing thought style, which almost always exerts an absolutely compulsive force upon his thinking and with which it is not possible to be at variance' (1979: 41).[1] According to Fleck, knowledge is created out of social stocks of knowledge:

Cognition is the most socially-conditioned activity of man, and knowledge is the paramount social creation. The very structure of language presents a compelling philosophy characteristic of that community, and even a single word can represent a complex theory. (Fleck, 1979: 42)

Fleck, therefore, asked 'To whom do these philosophies and theories belong?' His answer was that thoughts pass from one individual to another, each time a little transformed. 'Strictly speaking, the receiver never understands the thought exactly in the way that the transmitter intended it to be understood'.

---

[1] Thomas Kuhn (1970) popularized this idea in the 1960s. In the foreword to the English translation of Fleck's *Genesis and Development of a Scientific Fact*, Kuhn argues that when he first read Fleck's work, in 1949 or 1950, his German was so 'rusty' that he probably missed some of its points (Kuhn, 1979: p. ix). Indeed, Kuhn's idea of scientific revolutions is fundamentally different from Fleck's ecological idea of a population of simultaneous communities. Fleck also understood thought collectives as 'carriers' of a long history of accumulated knowledge and practice, not simply as competing views of the 'right' way to see the world. To use a distinction made by Latour and Woolgar, Kuhn's focus was the 'sociology of scientific knowledge' instead of the 'sociology of science' (Latour and Woolgar, 1986: postscript to the 2nd edn., p. 275).

After a series of such encounters, practically nothing is left of the original content. Whose thought is it that continues to circulate? It is one that obviously belongs not to any single individual but the collective. Whether an individual construes it as truth or error, understands it correctly or not, a set of findings meanders throughout the community, becoming polished, transformed, reinforced, or attenuated, while influencing other findings, concept formation, opinions, and habits of thought. After making several rounds within the community, a finding often returns considerably changed to its originator, who reconsiders it himself in quite a different light . . . The history of the Wasserman reaction will afford us the opportunity to describe such meandering in the particular case of completely 'empirical' finding.   (Fleck, 1979: 42–3)

Fleck argued that each thought collective has its associated practices. For every trade, every religious community, and every field of knowledge, there is a corresponding period of apprenticeship, during which a purely authoritarian suggestion of ideas takes place. In this process, novices gradually become experts. According to Fleck, every didactic introduction is therefore literally a 'leading into'.

The initiation into any thought style, which also includes the introduction to science, is epistemologically analogous to the initiations we know from ethnology and the history of civilization. Their effect is not merely formal. The Holy Ghost as it were descends upon the novice, who will now be able to see what has hitherto been invisible to him. Such is the result of the assimilation of a thought style.   (Fleck, 1979: 104)

As a result of this process, thought communities easily accept those new ideas that fit with its thought style. Incompatible ideas, in turn, are ignored and rejected as trifling and meaningless.[2]

According to Fleck, the general structure of a thought collective consists of a small esoteric circle and a larger exoteric circle, 'each consisting of members belonging to the thought collective and forming around any work of mind, such as dogma of faith, a scientific idea, or an artistic musing' (1979: 105). Furthermore:

A thought collective consists of many such intersecting circles. Any individual may belong to several exoteric circles but probably only to a few, if any, esoteric circles. There is a graduated hierarchy of initiates, and many threads connecting the various grades as well as the various circles.   (Fleck, 1979: 105)

Fleck also argued that in order to maintain their stability, thought collectives have to reproduce the concepts and facts that are central to their thought style, as well as their social structure. The new knowledge and practice that is created in the thought collective is applied in the context of a larger social system. Therefore there is also a constant need to negotiate the interactions between the esoteric and exoteric circle, which applies the products of the esoteric circle in social life. In this process, the esoteric circle of the community has continuously to legitimize and reproduce itself as the 'elite' of the thought community. Similarly, the thought community has continuously to negotiate and legitimize its position in the system of division of labour. If it succeeds in this, its knowledge and practices become institutionalized and its truths appear increasingly certain.

[2] This point was also made by Halbwachs (1980) and picked up by Mary Douglas (1987; 1996).

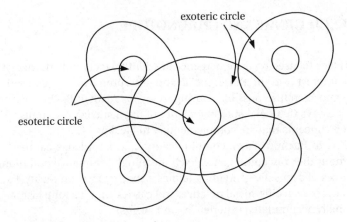

exoteric circle

esoteric circle

Fig.6.1. The general structure of thought collectives

Fleck argued that intercollective communication of ideas always results in a shift in their meaning. Whereas communication within a thought collective tends to filter out the novelty of ideas, and fit them with the existing way of thinking, when other communities appropriate ideas they always acquire new meaning. Words, for example, constitute a special medium of intercollective communication. But their meaning is always coloured by the thought style of the community that uses them. When words are used in a different community their meaning therefore changes. For example, the words 'force' and 'energy' mean different things to physicists, philologists, and sportsmen. In this process, the meaning of words may acquire some new colouration, change radically, or become completely destroyed. (1979: 109–10).

Fleck maintained that the force of thought communities easily overpowers any rationality or logical construction of individual thinking. A single individual can therefore—without much difficulty—be a member of several communities, and hold several 'logically incompatible' views of the world. This is not a problem for an individual as these contradictory views do not usually reach the stage of a psychological contradiction. The different realms of different thought communities are more or less independent. An individual usually participates in thought communities that are not closely related to each other. If thought styles are very different, their isolation can be preserved even in one and the same person. But if they are closely related, such isolation is difficult:

The conflict between closely allied thought styles makes their coexistence within the individual impossible and sentences the person involved either to lack of productivity or to the creation of a special style on the borderline of a field. This incompatibility between allied thought styles within an individual has nothing to do with the delineation of the problems toward which such thinking is directed. Very different thought styles are used for one and the same problem more often than are very closely related ones.   (Fleck, 1979: 110–11)

6.2 SPEECH GENRE AND CHRONOTOPE

Whereas Fleck focused on the emergence of scientific facts and knowledge, Mikhail Bakhtin's interest was in those social and interpretative processes that make language and literature possible. During the 1930s Mikhail Bakhtin wrote several important essays that developed the theory of linguistic genres. In his writings he introduced concepts that are complementary to Fleck's.

According to Bakhtin (1987), communication and knowledge cannot be understood without the concept of genre. Each meaningful utterance and thought uses the resources of a specific genre that defines a specific view on reality. Each genre, in turn, reflects a specific social practice, and carries with it a long history of trans-generationally accumulated experiences and events.

Bakhtin maintained that a competent individual has a large repertory of genres that can be flexibly utilized according to the situation. Everyday speech with family members, workplace conversations, discourses within academic disciplines, official letters, and poetic texts all require different genres. Without understanding the genre in use, meaning cannot be understood, and dialogue is impossible.

Bakhtin and his colleagues argued that the Russian formalists had it fundamentally wrong when they believed that the meaning of a text can be understood by synthesizing it from its constituent components. Also the semiotic view, which assumed that the users of language 'encode' meanings in words, was wrong. Instead, the meaning of words depends on the way they are used in a genre, and through the roles the words play within the associated social practice. Each word carries with it a 'halo' of meaning that reflects the way the world is seen from within a specific genre. Without this tacit halo of meaning, words are empty. Bakhtin maintained that genres are produced in a long historical process, tightly bound with evolution of specific social practices, and therefore the analysis of literature and language needs to be based on the analysis of genres.

Knowing, therefore, is about understanding the meaning within a specific genre and understanding what is true from the perspective of other genres. Knowledge is deeply rooted in culturally developed social practices and their trans-generational history.

As each genre has its own eye on the world and its own knowledge, dialogue across genres becomes a creative force. Bakhtin argued that the greatness of creative authors, such as Shakespeare and Dostoevsky, was in their ability to use the hidden potential of genres, and mobilize cultural resources that provide rich potential for interpretation. A great author does not only have a 'story to tell'. Instead, according to Bakhtin, creative authors have two different intents: they want to tell a story, but they also want to tell a story that does not remain confined within the limits of a specific story. Great authors compose a story with rich and open interpretation. They are able to do this because the genres themselves are open and constantly evolving, as reflections of the underlying social practices and history.

By intuitively relying on the cultural resources of a genre, a creative author can transcend his or her individual limits. According to Morson and Emerson (1990), an important objective of Bakhtin's work was to show that creativity is real, ongoing, and immanent in the process of daily living:

Thus, he sought to describe creativity without resorting to inspiration from the muse, intimations of the transcendental, or eruptions from the unconscious. Those models are usually used to explain the fact, noted at least since Plato, that poets seem to know more than they 'know'—to place more meaning in their works than they themselves could ever adduce. Without resorting to the irrational or mystical, Bakhtin's theory of genre accounts for this 'surplus' knowledge. Bakhtin is careful neither to explain away that surplus knowledge nor to call it inexplicable. Rather, it is the understandable consequence of prosaic creativity at work across cultures and through ages of great time. (Morson and Emerson, 1990: 298–9)

According to Bakhtin, genres evolve, interact, and become mixed. In this process new ways to understand the world and new social practices emerge. If we want to grasp the variety of ways in which the relation of people to their world may be understood, we need to examine the many concrete and detailed possibilities that literary genres have worked out (Morson and Emerson, 1990: 366). Bakhtin calls these concrete possibilities, which define the 'form-shaping ideology' of a genre, a *chronotope*. Each mode of thought is associated with its own genre and has its own chronotope. According to Bakhtin, actions are always performed in a specific context and chronotopes differ according to the way they interpret context and the relation of actions and events to it.

We will give the name chronotope (literally, 'time space') to the intrinsic connectedness of temporal and spatial relationships that are artistically expressed in literature. This term is used in mathematics, and was introduced as part of the theory of relativity (Einstein). (quoted in Morson and Emerson, 1990: 367)

Morson and Emerson note that the comparison between the Einsteinian time-space and Bakhtin's chronotope is important for several reasons. First, it implies that time and space are intrinsically interconnected, and events in a narrative cannot be understood without simultaneously understanding their place in time and space. Second, there is no 'objective' chronotope that would underlie and ground the specific chronotopes of different genres. Indeed, Bakhtin argued that Dostoevsky's creation of the polyphonic novel constituted a Copernican revolution: instead of using a single genre around which thought could revolve, Dostoevsky's polyphony generated a new space of interacting genres. This Copernican revolution, therefore, was bound to extend beyond its original limits and area of application. As Morson and Emerson note:

Indeed, all new genres, if they are based on profound enough forms of artistic thinking, have this potential for wide application, as do great philosophical or scientific theories. A new genre enriches our repertoire of visions of the world and may prove useful at distant times for unforeseeable applications, much as the non-Euclidian geometry of Lobachevsky and others found an unexpected application in relativistic models of the universe. (Morson and Emerson, 1990: 281)

Different aspects of the world also have their own rhythms and time-scales. For example, biological time-scales are different from astronomical time-scales.

Different social activities have their different chronotopes. The rhythms and spatial organization of 'the assembly line, agricultural labour, sexual intercourse, and parlour conversations differ markedly' (Morson and Emerson, 1990: 368). Chronotopes, then, become the ground for activity. They are not represented in the world. Instead, they are the ground which makes representability of events and social practice possible in the first place.

6.3 COMMUNITIES OF PRACTICE

For Fleck, a scientific fact was an expression of an underlying network of social interactions, technologies, and meanings. Bakhtin, in turn, emphasized the variety of the different systems of meaning, and their roots in history and social practice. Thought styles and speech genres are both products of social production and reproduction that cannot be reduced to the acts of individual actors or to society-wide processes. Instead, the level of analysis is between the individual level and the level of society. In other words, the foundation of meaning creation can be found from a 'community'. As the meanings of this community relate to its reality, and its reality relates to its practices, such a community can be characterized as a community of practice.

   Edward Constant presented an ideal model of a community of practice in his study on the history of the turbojet (Constant, 1980). He later developed this concept, arguing that community of practice is the locus of technological knowledge (Constant, 1984; 1987). According to Constant, members of a community of practice can be both individuals and organizations. In most cases, the membership is not defined by any existing technical discipline. For example, turbojet design utilizes precepts and people from aerodynamics, mechanical engineering, combustion engineering, metallurgy, and so forth:

What distinguishes a turbojet practitioner is adherence to the tradition, not disciplinary training . . . Turbojets are designed by a collection of engineers and other specialists, who together constitute an identifiable community of practitioners . . . The normal practice of such communities, however they are defined, is the extension and articulation, or incremental development, of the received tradition.   (Constant, 1987: 224–5)

Although the normal pattern of development in such a community of practice is incremental, sometimes the community may face persistent failure, which may provoke a search for radical new solutions. Sometimes it is also possible that new theoretical insights may imply that a current system will fail under future conditions, and that a radically different new system may work better. As we noted before, Constant called such situations *presumptive anomalies*. For example, in the 1920s aerodynamic theory suggested that with sufficient thrust a well-streamlined aircraft should be able to approach the speed of sound. Theory also showed that conventional propellers could not operate at such speeds. Theory, however, also indicated that gas turbines could provide the required thrust. The presumptive anomaly, therefore, led to the turbojet revolution.

Constant argued that communities of technological practice embody higher-level traditions of testability. These traditions of testability define the norms, criteria, and generic approaches which are followed during incremental change in a community. For example, engineering communities normally adhere to generally accepted criteria on how to measure the success of a design. When change is radical, the standards of testability, as well as the criteria for success may, however, change. For example, the turbojet revolution required entirely new testing techniques and testing facilities. It also required redefinition of the way engine output is measured. Whereas aircraft engines had earlier been measured by their shaft horsepower, turbojets were measured by their thrust. According to Constant:

Traditions of technological testability, then, linked both to lower-level traditions of technological practice and to a higher-level normative engineering culture, are the major way that communities of practitioners reify the meaning of their tradition of practice for themselves and explain and justify that tradition to outsiders.   (Constant, 1987: 226–7)

Constant argued that complex technological systems can be hierarchically decomposed into subsystems, and that this implies multiple traditions of practice and multiple communities of practitioners. A technical change that is incremental development of the complex system may therefore be radical when it is perceived at a level of a subsystem. According to Constant, each subsystem can be seen as the purview of a distinct community. There can be overlap between communities at the different levels of the hierarchy, as well as between the various communities at the same hierarchical level.

Constant's discussion on communities of practice emphasized the link between technological architecture and community structure. The division of labour that was required to effectively create a complex technological system created a system of interdependent communities. Moreover, there was a duality between the system and communities related to it: social structure and technological structure reflected each other, and change in either of them required adjustment in the other.

6.4 SOCIAL LEARNING IN COMMUNITIES OF PRACTICE

Much of the recent literature in knowledge management has emphasized the importance of social learning in understanding how knowledge is created and located in social systems. For many authors, a central idea has been Vygotsky's model of social learning, and his concept of a zone of proximal development (e.g. Engeström, 1987; 1999; Spender, 1996; 1998; Brown, Collins, and Duguid, 1989; Wenger, 1998; Nardi, Whittaker, and Schwartz, 2000; Tuomi, 1998). The basic idea in Vygotsky's theories was that knowledge is produced in a socio-historical process, and that it is tightly linked to social practice (Cole, 1986; 1996; Wertsch, 1985; 1998; van der Veer and Valsiner, 1994; Kozulin, 1990; Daniels, 1996).

Vygotsky argued that three different lines of development interact in cultural-historical evolution. In addition to the evolution of species, culture accumulates concepts and practices, which are, in turn, appropriated by an individual during his or her individual development (Luria and Vygotsky, 1992). Vygotsky therefore distinguished evolution of phenotype, culture, and individual ontogenic development. Ontogenic development, however, is not independent of social interaction and culture. According to Vygotsky, important social stocks of knowledge, such as language, are learned in a process where members of the society support the enculturation of a developing child. The zone of proximal development was defined by Vygotsky as that range of possible action and thinking in which an individual could perform with support from his environment (Vygotsky, 1978: 84–91; van der Veer and Valsiner, 1994: 336; Rogoff, 1990; Tuomi, 1999: 315). By moving in this zone, the developing individual can expand his or her competences, finally becoming a fully competent adult.

Jean Lave and Etienne Wenger applied this Vygotskian concept of the zone of proximal development in a cultural-anthropological context (Lave and Wenger, 1991). They argued that the locus of expertise is in a community of practice, and novices become members of the community through gradual social learning. This, however, requires that they get legitimate access to the community in question. Lave and Wenger introduced the concept of *legitimate peripheral participation* to describe the starting point of such social learning. Furthermore, they argued that knowledge is bound with the practices and tools used in the community. Lave and Wenger therefore focused on learning as a process that socializes individuals as members of traditions of practice.

Yrjö Engeström (1987) relied on the Vygotskian cultural-historical tradition and its ideas on social learning, but his focus was on the question of how communities create new practices. Based on A. N. Leont'ev's (1978) theoretical model of the structure of social activity and Vygotskian ideas that provided the foundation for Leont'ev's activity theory, Engeström developed a model of activity systems. In Engeström's model, activity is defined as symbol- or tool-mediated object-oriented activity, embedded in a community that operates under a given division of labour and rules that regulate it. Based on his model, Engeström further formulated a model of expansive learning, as a form of change that creates new forms of activity. According to Engeström, new forms of activity are generated by contradictions in and between existing systems of activity. For example, when one activity system produces tools that are used in another activity system, change in one of these systems may trigger learning processes in the other.

In other words, in Engeström's theory an activity system consists of tool-mediated activity that occurs in the context of a division of labour, community of practitioners, and institutions that regulate exchanges. The locus of innovation is, therefore, a complete activity system. Learning of new practices occurs in a cycle where contradictions are detected and interpreted, and possible solutions are generated, tested, and finally integrated into current practice.

The actual process of learning in Engeström's model (Engeström, 1987; 1999) resembles Dewey's (1991) cycle of experimental learning and Schön's (1983) reflective practice.[3] As the basis of Engeström's model is in cultural-historical activity

---

[3] I have compared these learning models in Tuomi (1999: 306–40).

theory, his model, however, is fundamentally about social learning. In this model, the sources of new innovative practice are inherently social.

6.5 THE CONCEPT OF *ba*

Nonaka (1994; Nonaka and Takeuchi, 1995) proposed an influential model of organizational knowledge creation based on a distinction between tacit and explicit knowledge. According to Nonaka, knowledge is created in a process where tacit knowledge is socially shared and converted into explicit knowledge, and explicit knowledge is combined with existing explicit knowledge and converted back into tacit form where it guides action. Nonaka called these different modes of knowledge conversion 'socialization', 'externalization', 'combination', and 'internalization', and the resulting knowledge creation model the SECI-model.

In the SECI-model, knowledge 'spirals' from the individual level to the level of teams, organizations, and beyond the organization. New knowledge is cross-levelled and converted in this process. Cross-levelling, however, is not simply a unidirectional flow of knowledge from individual innovators to society; instead, there is constant interaction between the different levels of analysis. Individuals are members of organizations and teams, and knowledge creation occurs simultaneously at the individual and social level. Knowledge emerges in social interaction.

By making the distinction between tacit and explicit knowledge, Nonaka was able to represent the SECI-model in a compact way. This basic SECI-model is shown in Fig. 6.2.

Based on the SECI-model of knowledge creation, Nonaka and Konno developed a model of knowledge creation spaces. They proposed that each quadrant in the SECI-model has an associated 'space' of knowledge creation. Using the Japanese concept *ba* that roughly translates as 'space' they proposed that there were four different types of *ba*s.

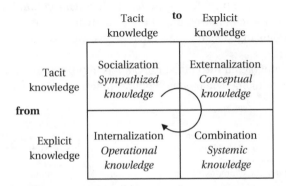

Fig. 6.2. The basic SECI-model of knowledge creation

Ba is a sophisticated concept and cannot simply be understood as physical location or space. According to Nonaka and Konno (1998), *ba* can be thought of as a shared space for emerging relationships. This shared space provides a platform for advancing individual and collective knowledge. *Ba* is a context that harbours meaning, and which acts as a medium where individual knowing transcends its limits, thus enabling the creation of new knowledge.[4] Nonaka and Konno argued that if knowledge is separated from its *ba*, it becomes information, which can then be communicated independently from *ba*.[5]

According to Nonaka and Konno, tacit-to-tacit conversion occurs in *originating ba*, tacit-to-explicit conversion in *interacting ba*, explicit-to-explicit in *cyber ba*, and explicit-to-tacit in *exercising ba*. The originating *ba* is the space of social interaction where new knowledge emerges. In the interacting *ba* individuals' mental models and skills are converted into common terms and concepts. The interacting *ba*, which Nonaka and his colleagues later called 'dialoguing *ba*' (Nonaka, Toyama, and Konno, 2000), is more consciously constructed than the originating *ba*, and often organized as cross-functional teams and task forces. The third *ba*, 'cyber *ba*' or 'systemizing *ba*', represents the combinatory phase in knowledge creation. This *ba* combines existing explicit knowledge and information. According to Nonaka and his colleagues, a systemizing *ba* offers a context for combination of existing explicit knowledge, and technologies such as email and information systems can be used to support such combination processes. Whereas originating *ba* and dialoguing *ba* are defined primarily by face-to-face social interaction, systemizing *ba* can rely on knowledge artefacts, such as documents, email messages, databases, and computer-mediated collaboration.

The fourth type of *ba*, exercising *ba*, is the context of internalization. In this *ba*, 'individuals embody explicit knowledge that is communicated through virtual media, such as written manuals or simulation programs'. According to Nonaka *et al.*, in contrast to dialoguing *ba* where the limits of current knowledge are transcended by thought, in the exercising *ba* these limits are transcended through action (Nonaka, Toyama, and Konno, 2000).

According to Nonaka and his colleagues, *ba* exists on many ontological levels. Individuals form the *ba* of teams, which in turn form the *ba* of organization. The market environment, in turn, forms the *ba* for the organization. Innovation management, and strategic management of the firm, therefore, becomes a problem of leading and organizing *ba*s and their interactions (Nonaka, Toyama, and Nagata, 2000).

To properly understand the concept of *ba*, it is important to note its central role in the Japanese thinking. The Japanese philosophical tradition of the Kyoto School

---

[4] 'Transcending the limits' should here be read in the phenomenological sense of 'transcendence': accessing the world beyond what is currently included in the world of meaning, and therefore accessible to a conscious cognition.

[5] This is a problematic claim, as I have argued in more detail in Tuomi (2000). Also the SECI-model itself is problematic, for example, as the different knowledge conversion modes and levels of analysis are not easy to separate conceptually or in practice (cf. Tuomi, 1999: 326–40).

developed the concept of *ba* to discuss the fundamental problems of meaning creation and the nature of objects and knowing subjects. Kitaro Nishida (1870–1945), the founder of the Kyoto School, studied Western philosophers in great detail, and integrated insights from Western phenomenological thinkers to insights of thinkers in the Buddhist tradition. In his different works, Nishida approached the problems of meaning creation through William James, Henri Bergson, and Edmund Husserl, among others, often providing penetrating analyses of these thinkers.[6]

A central concept in Nishida's later philosophical thinking was *basho*. The concept of *basho* had its roots in the idea of topos, which Plato discussed in his *Timaeus*, in Aristotle's notion of hypokeimenon, and in Lask's field theory. The modern physical concept of field of force also played an important role (Van Bragt, 1982: pp. xxx–xxxi). In the context of Nishida, *basho* is most often translated as 'locus' or 'topos'. Nishida's student Nishitani used the term *ba* as a philosophically central concept to describe the field of meaning creation. The use of the field concepts enabled the Kyoto School philosophers to overcome the conventional Western subject- and object-centric notions of knowledge and knowing.[7]

Although *ba* can be translated as 'space', it is important to note that the concept of *ba* is based on an epistemology and ontology that resulted from a critical analysis of empiristic and objectivistic theories of knowledge. Nishida tried to develop a logic that could describe the mutual construction of subjects and objects. This led him to reject the Aristotelian logic of assertions and the Hegelian logic of contradictions. Instead of the Hegelian dialectic logic, where thesis and antithesis produce synthesis, Nishida argued that we need a logic of 'contradictory identity' that enables us to describe phenomena that may look paradoxical in the context of Aristotelian logic. For example, the world produces us as living subjects of the world and therefore we are part of the world. As conscious and acting subjects, we, however, also constitute the objects of the world as meaningful phenomena of the world. In other words, we are 'expressions' of a world, which, in turn, is expressed in our interaction with it. In Nishida's terms, when we interact with the world our consciousness creates the world as the object of our action, but at the same time we are created as actors in that world. In a very concise, but somewhat cryptic way, Nishida summarizes this by saying that we think and act by becoming things (1987*b*: 55). The fundamental starting point of most Western philosophies was the independence of subjects and objects, and the logic of 'is' or 'is-not', and therefore Nishida argued

---

[6] Although his works are not widely known among Western philosophers, Nishida is one of the most sophisticated philosophers of the phenomenological tradition. His writings have been translated into English in Nishida (1958; 1987*a*; 1987*b*; 1990). Nishida's epistemology has some similarity with the process philosophy of Alfred North Whitehead and the existential phenomenology of Emmanuel Levinas. See also Nishitani (1991), Carter (1997), Takeuchi (1963), Axtell (1991), and Nishida (1920).

[7] In this the fields of *ba* and *basho* are similar to modern quantum field theory, where objects become manifestations of underlying fields. Quantum phenomena cannot be understood through the analysis of these objects themselves; instead, such understanding requires analysis of the fields that express themselves in objects under suitable boundary conditions. And, as Einstein showed, 'objects', in turn define the spatio-temporal structure where they exist. *Basho*, therefore, is also similar to Bakhtin's chronotope.

that we need a new foundation for philosophical thinking. Therefore, the correct starting point for a theory of knowing is a logic that doesn't separate the knower and the known, but which sees them as mutually constructed. In a somewhat simplified way, one could say that the *ba* is the 'space' where this mutual construction transpires.

The 'field' of knowledge creation, therefore, cannot be reduced to a physical space or location in any straightforward way. Perhaps it could be best conceptualized as a dynamic 'space of mutual construction'. The space and its objects are actively created in the human mind in interaction with a world that gradually becomes a world of meaningful objects, in the process of individual and cultural development. *Ba*, therefore, exists on a level where meaning emerges, and it is a more fundamental construct than 'space' and its 'objects'. In Polanyi's (1998; 1967) terminology, *ba*, therefore, includes the 'tacit dimension' of knowing. This view, of course, is in stark contrast with most conceptualizations that underlie the discussions of information processing, knowledge sharing, and decision-making in organizations.[8] In most Western epistemologies, the basic assumption is that true knowledge has to be independent of the knower, and that knowledge emerges when the objects and phenomena of the world reveal their true nature.

The way Nonaka and his colleagues (Nonaka and Konno, 1998; Nonaka, Toyama, and Konno, 2000; von Krogh, Ichijo, and Nonaka, 2000) used the concept of *ba* has its roots in the Kyoto School and in Polanyi's analysis of meaning fields. According to Polanyi, knowledge emerges from a field of meaning as this field becomes articulated as explicit knowledge. In the context of the Kyoto School, as well as in Polanyi's writings, explicit knowledge exists as a dimension of the field of meaning, and cannot be separated from it. Although many readers of Nonaka and Takeuchi (1995) have concluded that the SECI-model of knowledge creation is based on two different 'stocks of knowledge', this interpretation is therefore incompatible with a more exegetic reading of the underlying epistemic concepts.[9]

---

[8] It is therefore not a great surprise that it also leads to non-conventional views on the nature of organizations and their meaning processing. It also leads to computer designs that radically differ from conventional information-processing systems. Shimizu has used the concept of *ba* to develop his holonic computer, which aims at meaning processing, instead of information processing (Shimizu and Yamaguchi, 1987; Heinämaa and Tuomi, 1989: 270–1). Nonaka regards Shimizu as a thought leader behind the modern concept of *ba*.

[9] This error has been made, for example, by Brown and Duguid (2001), as well as myself (Tuomi, 1999; 2000), as we have argued that Nonaka's SECI-model doesn't take into account the fact that tacit and explicit components cannot be separated. Brown and Duguid make the point that the SECI-model is problematic as Polanyi's distinction between tacit and explicit knowledge requires that they are inseparable, but often the clear separation between tacit and explicit components of knowledge is understood to be the main contribution of the SECI-model. Many popular introductions, as well as many academic authors, assume that tacit and explicit knowledge are two different 'stocks' of knowledge, which can be independently stored and processed (cf. Tuomi, 2000). Nonaka's use of tacit–explicit distinction is somewhat ambiguous and it is not clear how closely he follows the thinking of the Kyoto School. It is not clear whether, for example, the idea of 'detaching knowledge from its underlying *ba* and thus turning it into information' would make sense for Kyoto philosophers. To say so in the context of the Kyoto School would require a sophisticated concept of information, which probably is not close to what most contemporary readers make of it.

As Bakhtin noted, concepts and words make sense only in the context of a genre and chronotope. For Bakhtin, creativity is located in interpretation but also in the creative use of cultural resources.

Nishida's *basho* has many similarities with Bakhtin's chronotope. Both describe the topos that underlies a meaningful world. For Nishida, however, *basho* underlies the meaning of the world in a fundamentally existentialist sense. Outside the space of *basho* there is nothing that has meaning. The universe of *basho* is closed, expanding, but without boundary. The solipsism of this universe is broken in a radical way when Nishida proposes a new paradoxical logic, arguing that our conventional logic implies a unique and misleading concept of reality where subjects and objects are separate. To understand the way we exist in the world, we therefore have to switch to a new genre built around a new logic of mutual construction.[10]

A detailed analysis of these rather profound epistemological and ontological ideas might reveal important theoretical questions that perhaps lead to new insights concerning the nature of *ba*s. But based on the discussion above we can already ask one such question. This is the question how social learning and knowledge-creation processes interact in *ba*s. In the next chapter I'll propose that *ba*s comprise two qualitatively different layers of knowledge articulation, and that there exist also two different innovation dynamics that make it necessary to separate the four types of *ba* proposed by Nonaka and his colleagues into two qualitatively different classes. These distinctive types of knowledge-creation spaces became visible when we discuss the ways in which the social infrastructure of knowing evolves.

---

[10] As Nishida's philosophy is based on analysing our way of being in the world, it has been categorized in the Western world as a philosophy of religion. The reason is partly that Nishida's thinking is related to the Buddhist tradition, and partly that he wrote extensively about what could be called the ethical consequences of his existentialist phenomenology. In this sense, Nishida comes quite close to Bergson (1977), Buber (2000) and Levinas (1969). Of course, any philosopher who proposes a radically different foundation for epistemology, ontology, and logic easily looks like a 'philosopher of religion' to those philosophers who practise philosophy in another genre.

7

Breaking through a Technological Frame

As we have seen above, there are many different and complementary ways to conceptualize practice-related communities and their knowledge-production processes. Although we now have a relatively rich theoretical understanding of the nature of communities of practice, we still have an interesting problem to solve: if new technology creates new social practice, innovation means that communities change. How does this happen, and what, exactly, do we mean by such change? Does the community of practice concept apply only to isolated traditional tribes and conservative communities that carefully protect their knowledge and practices, or can we use the concept also in the context of innovation?

To answer these questions it is necessary to discuss two different developmental dynamics that generate change in communities. Here it is useful to start by describing in more detail Bijker's concept of technological frames.

According to Bijker, a technological frame is composed of the concepts and techniques employed by a community in its problem solving:

Problem solving should be read as a broad concept, encompassing within it the recognition of what counts as a problem as well as the strategies available for solving the problems and the requirements a solution has to meet. This makes a technological frame into a combination of current theories, tacit knowledge, engineering practice (such as design methods and criteria), specialized testing procedures, goals, and handling and using practice. (Bijker, 1987: 168)

In the history of plastics, Adolf Bayer is often portrayed as the first researcher to produce a synthetic resin. According to these histories, after Bayer produced the synthetic resin, researchers directed their efforts toward rendering it in an industrial process. Eventually Leo Baekeland succeeded in this, and the synthetic plastic industry was born.

Bijker, however, noted that for Bayer the resin he produced was something completely different than it was for Baekeland. Bayer was looking for a synthetic dye, and the phenol-formaldehyde resin his experiments produced was an annoying by-product that had to be thrown away. Similarly, when scientists learned to synthesize cheap formaldehyde in 1888, the rapidly dropping costs of resins did not lead to the emergence of synthetic plastics. Although researchers studied formaldehyde reactions they did it to find raw material for the synthetic dye industry.

Their experiments failed, as formaldehyde reactions often generated resins that spoiled the experiment. For example, an industrial chemist Werner Kleeberg tried to analyse the produced resins, but after failing in this, focused on reactions that did not produce them (Bijker, 1987: 167). Bijker notes:

If we now apply the concept of a technological frame to the discussion of Bayer and Kleeberg, it becomes clear why they did not try to modify the phenol-formaldehyde condensation product into a usable plastic. First, they had other goals: the production of new synthetic dyes. But these goals can be changed, especially when large profits are on the horizon. So there is more to it than this. The idea of making plastic by chemical synthesis simply did not and indeed *could* not occur to them. Chemical theory at that time could not cope with such a substance. Neither could chemical practice: Their daily laboratory practice included all kinds of chemical analysis and synthesis, but the application of pressure and molding techniques were of another world. The technical frame of synthetic plastics was not yet in existence.   (Bijker, 1987: 168–9)

Where did such a frame, then, come from? According to Bijker, Baekeland started his work in the frame that had its origins in Celluloid production. Celluloid was developed to replace ivory in billiard balls, using nitrocellulose plastics as a starting point. The first synthetic plastic was developed by Parkes in 1865, first to address the perceived imminent scarcity of the supplies of India rubber and gutta percha, and then to substitute ivory. Parkes and his Parkesine did not succeed in business terms, but several attempts were made to correct the early failures. John Wesley Hyatt first tried to make ivory-like substance from nitrocellulose in the late 1860s, but quickly noted that the semi-liquid nitrocellulose could not be used as the drying process inevitably caused shrinking. After some experimentation, he found a way to use camphor, heat, and pressure to produce Celluloid.

Celluloid was a successful product and its use become stabilized among several communities of users. It did, however, also have important problems. For some uses its flammability was a problem, and the high cost of camphor made it expensive. Researchers therefore tried to invent a substitute for it. No successful substitutes were found.

Only when Baekeland rejected the approaches used by the plastics researchers, Bakelite was born. When he failed to find a suitable production process, he switched to techniques used in electrochemical engineering. After carefully studying the chemical reactions of phenol-formaldehyde condensation, Baekeland learned that it had three different stages. Furthermore, he also learned how to stop the process before it moved to the next stage. As a result, Baekeland was able to develop a thermosetting plastic that could be moulded and used to produce a strong synthetic material. At the same time this new material produced a new understanding of what plastics are and for what purposes they could be used.

Bakelite was rapidly taken into use by several industries, and a community of Bakelite practitioners was formed. The technological frame of Bakelite emerged in this interaction between users and producers.

A technological frame is built up when interaction 'around' an artefact starts and continues. Thus the artefact Parkesine did not give rise to a specific technological frame because the interactions 'around' it came to an end before really taking off. The opposite happened to Celluloid: Its stabilization was accompanied by the establishment of, for example, a social group of 'Celluloid chemists'. The continuing interactions of these chemists gave rise to *and* were structured by a new technological frame.　(Bijker, 1987: 173)

But although a technological frame structures interactions of members of a social group, it never completely determines them. This is because the members have different degrees of inclusion in the frame, and because actors are generally members of several technological frames. The inclusion itself has a multidimensional character: technological frames consist of goals, problem-solving strategies, experimental skills, and theoretical training, and therefore frames are multifaceted. Although, for example, Baekeland's goals were congruent with the Celluloid producers' technological frame in that Baekeland tried to produce plastic articles, his goals were also different as he was trying to produce industrial applications, instead of consumer goods (Bijker, 1987: 174).

Baekeland first assumed that he could issue licences for plastic manufacturers. Soon he realized, however, that he could not do it. The concepts, practices, and skills that were needed for Bakelite production did not exist within the Celluloid framework:

I found, to my astonishment, that people who were proficient in the manipulation of rubber, celluloid or other plastics were the least disposed to master the new method which I tried to teach them or to appreciate their advantages. This was principally due to the fact that these methods and the properties of the new material were so different in their very essence from any of the older processes in which these people had become skilled. This rather unexpected drawback is so true that even to-day the most successful users of bakelite are just those who were not engaged in plastics before . . .　(Baekeland, 1916; quoted in Bijker, 1987: 175–6)

The early community of Bakelite practitioners consisted of the employees of the Bakelite Corporation. Competing plastic producers tried to void Baekeland's patents, but after they failed they rapidly learned to understand plastics according to the new frame. Baekeland was also able to enroll engineers from the new electronics industry into his technological frame. Through the electronics industry Bakelite moved to the automobile industry, which was looking for a strong oil-resistant material that could be used as an insulator. This stabilized the community of Bakelite production, and by the end of the 1930s several new groups enrolled to the frame as users (Bijker, 1987: 176–7).

Using his analysis of the evolution of technological frames, Bijker argued that innovation, in general, occurs in three stages or settings. First there is no community whose technological frame would be dominant. For example, when the bicycle emerged around 1880, its potential was understood in very different ways by different social groups, and no single group was able to define what a 'bicycle' is. In the second stage, a single social group becomes dominant and the innovation is interpreted using its technological frame. For example, for a considerable time, plastics were understood in the context of the Celluloid frame. In the third stage, new social groups start to use their distinctive technological frames to interpret the innovation.

According to Bijker, in the first stage there is not much constraint in the ways an innovation is interpreted. Therefore, many radically different variants emerge. For example, around 1880, when safety was seen as the major problem for bicycles, many different constructions were produced. In the American Star bicycle a small steering wheel was positioned ahead of the high wheel; Lawson's Bicyclette had a chain drive on the smaller rear wheel; and, in general, all aspects of the design were subject to variation, leading for example to three- and four-wheel 'bicycles' (Bijker, 1987: 182–3; 1997: 19–100).[1]

Bijker argued that in this state one important way to stabilize a technological frame is to enrol existing groups into the new frame. One way to do this is to redefine the technology as a solution to a problem for the group that is to be enrolled. For example, when air tyres were first introduced for bicycles they were offered as solutions to the problem of vibration. As this failed, they were offered as solutions to a speed problem. As this problem was important to racing cyclists, this user group was enrolled, and air tyres were taken seriously.

According to Bijker, when one technological frame is dominant, it is possible that technology has a *functional failure*. Functional failure is a source of innovation especially for people who are in the core of the community. Functional failure may occur when an existing technology is used under new and more demanding conditions.[2] Bijker argued that functional failure usually leads to incremental innovation, as problems are addressed within the existing technological frame. When Celluloid producers realized that flammability was a problem in many uses of the technology, they made efforts to find a solvent that could be used to produce a less flammable 'Celluloid'.

Actors who do not have high inclusion in a technological frame have less difficulty in using problem-solving strategies that are not part of the frame. Whereas high-inclusion members of the community tend to be sensitive to functional failures, low-inclusion members tend to see presumptive anomalies. In other words, low-inclusion members can see the technology being used under radically new conditions, and therefore also imagine situations where a currently working technology would fail. As was noted above, such a presumptive anomaly was seen, for

---

[1] It has often been noted that the early phases of development of a new technology create a large variety of product designs. Utterback and Abernathy (1976; Utterback, 1994) argued that a 'dominant design' emerges when key aspects of the new technology become standards for the product. Bijker's analysis turns this view around by looking at the stabilization of a technological frame. Instead of a dominant design, as a list of product features, we therefore have a dominant technological frame which connects the product to the various practices where the product makes sense. This interpretation also makes it understandable why high-tech competition seems to be inter-industry competition and why dominant designs don't seem to stabilize in high-tech (Kodama, 1995: 100). We simply tend to define as 'high-tech' products that are essentially combinatorial.

[2] Petroski (1994; 1996) showed that this is a common occurrence in the evolution of technical designs. He studied the evolution of designs for such everyday objects as forks, zippers, paperclips, and pencils, and argued that 'form follows failure'. For example, when the fork became used for putting food in the mouth in 16th- and 17th-century. Europe, the traditional forms of fork failed. The fork changed from a 'tool-for-holding-meat-while-cutting' into a 'tool-for-eating-without-touching-the-food'. Both its design and meaning changed accordingly.

example, by aerodynamic researchers who realized that propeller engines could not work at speeds that were close to the speed of sound. Bijker maintained that especially young, recently trained engineers are in a position to recognize such presumptive anomalies, as they are trained within a technological frame, but have low enough inclusion to question the basic assumptions of the frame (Bijker, 1987: 184).

According to Bijker, a third stage of innovation occurs, and a new innovative setting emerges, when several competing technological frames exist. For example, at the end of the nineteenth century, electric power networks were built based both on alternating current and direct current. Often both systems were used in the same town. According to Bijker:

The selection process in a situation like this is quite hectic, more so than in the first situation, in which there is no dominant technological frame and when less vested interests are at stake. Arguments, criteria, and considerations that are valid in one technological frame will not carry much weight in other frames. In such circumstances it seems that criteria that are external to both technological frames will play an important role in the selection process. This makes rhetoric a fitting selection mechanism in this third situation.   (Bijker, 1987: 184)

For example, Thomas Hughes (1983) documented such a rhetorical move in the 'battle of currents'. A dog was publicly electrocuted by subjecting it to various voltages of alternating and direct current. The objective was to persuade the audience that alternating current was relatively safe.

Bijker noted that these three different situations can be found by studying the history of technology, but that they are not always easy to separate. Indeed, although Bijker's description of the role of technological frames is useful and insightful, it does not seem to explain why the three different stages can sometimes easily be detected, and sometimes not.

Part of this problem is due to retrospection. For example, studies on discontinuous and disruptive technological change have described technological evolution as a process that is simultaneously continuous and discontinuous (Tushman and Anderson, 1986; Anderson and Tushman, 1990; Henderson and Clark, 1990; Rosenbloom and Christensen, 1994). For example, computer hard disk drive technology has undergone several changes where new uses have become dominant (Christensen, 1997). At each transition, leading producers have lost their leading position and new entrants have become main players in the industry. Yet, we can describe such disruptive change only because it is framed in a context of an 'industry' where each new generation of technology is seen as a next step in the continuous evolution of 'disk drive technology'. In retrospection, it would also be easy to argue that Bakelite generated a disruptive change in the plastics industry. As Bijker noted, however, the uses and interpretation of plastics changed as a result of the emergence of a new Bakelite practice.

As this example shows, discontinuity is not associated with characteristics of technology as much as it is associated with our interpretations. It may take some time to draw a picture that can be seen either as a rabbit or a duck, but we may change our interpretation of the picture in a fraction of a second. Although nature itself may not make discontinuous jumps, our interpretations of reality may do so.

The famous rabbit-duck, however, is a rather typical laboratory animal. It exists because a picture of a rabbit is not a rabbit. To see something as a rabbit means that this something acts like a rabbit and makes similar differences to our life as a rabbit does. In most cases we see rabbits as rabbits and ducks as ducks because we see them in ecological environments where they make sense (Gibson, 1979). Disruptive technological change, therefore, reflects change in social practices. When new practices become dominant, the old framework doesn't work anymore.

Where, then, does this change come from? To understand this, we have to distinguish two different processes that generate new technological frames. These two different processes lead to two different modes of innovation. The first has its source in functional differentiation, specialization, and division of labour in social systems. The second becomes possible when functional differentiation has created enough complexity in the system of social practices so that recombination becomes possible.

7.1 TWO EVOLUTIONARY PATHS OF COMMUNITIES

As was noted earlier, when the various authors discuss communities and practice, the emphasis varies from social learning to collaborative production and generation of new forms of practice and knowledge. These different aspects of community life cannot easily be separated. An important distinction can, however, be made between *homogeneous communities*, where members essentially share the same interpretation of the world, for example, a community of Yucatan midwives, insurance claims processors, or Linux operating system developers, and *heterogeneous communities* where members speak different languages and have different tacit stocks of knowledge. A typical heterogeneous community can be formed around laboratory practice, for example, around an operating room in a hospital, or around a technological artefact, such as a jet engine or a new communication computer.

Another important distinction can be made based on the stability of these communities. Some communities are ephemeral, others are stable, and some become institutionalized. In heterogeneous ephemeral communities the members don't have much time to learn each other's language, whereas in institutionalized heterogeneous communities new homogeneous communities can emerge on top of the existing social structure. Together these two dimensions of heterogeneity and stability to a large extent determine how knowledge can be created and exploited in such communities (Tuomi, 1999: 272–5).

Why, then, are some communities homogeneous whereas others are heterogeneous? How are these communities generated? I shall explore these questions below. My argument will be that two different developmental dynamics lead to these different types of communities. At the same time, these two different developmental paths lead to two different ways these communities create innovations and new knowledge. One innovation dynamic is generated by an increasing division of labour and functional differentiation in social systems, and another by the combination of resources generated by existing communities.

7.2 DEVELOPMENT OF SPECIALIZATION, DIVISION OF LABOUR, AND NEW TECHNOLOGICAL FRAMES

As was noted above, social learning reproduces communities of practice and leads to incremental change in their stocks of knowledge. Communities can, however, also spin off new communities as a result of their developing division of labour. The evolution of communities can therefore produce an ecology of communities of practice, each with their social stocks of knowledge.

This theoretical insight was at the core of Marx's epistemology and sociological theory. Marx argued that society and the human mind have to be studied in the context of developing human practice. Indeed, Marx maintained that the human mind itself has been produced in a process where social practice has become differentiated. For Marx, collaborative and goal-oriented production was a specific characteristic of the human species, which made humans different from animals. This collaborative production he called labour. The emergence of labour required the emergence of collective coordination, communication, and meaning. Therefore, labour and the human mind had their source in the same process of division of labour.

Durkheim (1933) maintained that it may be an exaggeration to say that human psychic life starts only when societies emerge. He argued, however, that as long as the society consists of only few members or is not geographically concentrated, even the most developed forms of psychic life remain communal. In traditional societies, society equals community. In Durkheim's terms, traditional societies are based on 'mechanical solidarity', which has its roots in communally shared values and life practices. Mechanical solidarity, in other words, is based on similarities. The evolving division of labour, however, creates a new form of society where the different groups of the society become interdependent on each other's work. As a consequence, modern society is characterized by 'organic solidarity'. Durkheim argued that the actual forms of division of labour can only be understood when the value systems of the different collaborating groups are taken into account. Without some overlap between the collective consciousnesses of different groups and cultures, true division of labour could not be possible.

For Marx, the main characteristic of modern industrial society was that it was a capitalistic society. The most important social difference was to be found between those who owned the means of production and those who didn't. As a consequence, in Marx's thinking the important aspect of collective consciousness was class consciousness. The future was therefore built on changing the worker's mind. Underlying this view, however, was the conviction that the human mind is constituted through praxis, which is inherently social and has its roots in division of labour. The founders of cultural-historical activity theory, including Vygotsky, Luria, and Leont'ev, were well aware of this insight. Cultural-historical activity theory connected this idea with theoretical models of learning, activity, knowledge creation, semiotics, and technology use. This led to extensive theoretical and experimental work on the social basis of human cognition (cf. Wertsch, 1981; 1985; 1991;

van der Veer and Valsiner, 1994; Kozulin, 1990; Wertsch, del Río, and Alvarez, 1995; Engeström, Miettinen, and Punamäki, 1999).

Leont'ev (1978) illustrated the evolution of activity by analysing the development of fundamental human activities, such as hunting. In early phases of cultural development, hunting can be the simple catching of game. The object of this activity is the prey, and the motive of activity is hunger. Early on, the evolving division of labour, however, changes the picture. A group of hunters splits into two groups: one that beats the bushes to frighten the game, and another that waits silently for the frightened game to come close enough so that it can be killed. The action of noise makers and game killers is meaningful only in the context of the activity of hunting. The meaning of human activity, therefore, is inherently connected to the social division of labour (Leont'ev, 1978; Axel, 1997).

Leont'ev proposed a three-level model of human activity. *Activity* itself occurs as meaningful social productive practice. It has an object, such as food, that can also be understood as its motive.[3] Activity, in turn, is realized through *action*. We can observe activity only in actions and their sequences. The meaning of action can only be understood in the context of activity, and the same action can be used in realizing different activities. For example, we can beat bushes for many different reasons. Therefore we cannot infer activity from actions. They exist in different worlds that cannot be reduced to each other.

The third level in Leont'ev's model is the level of *operations*. Actions are implemented on this level. The way actions are made concrete depends on the concrete situation at hand. For example, hunting may be realized through hiding behind a tree until the game is close enough, but the way the hunter does this depends on the trees and tools used for killing the animal.

According to Leont'ev, these three fundamentally independent levels of activity are in constant movement. For example, the noise makers can develop better tools to make a louder noise, and eventually noise-making can become an end in itself. Part of the group of hunters can specialize in drumming, which can evolve into activity with the object of making music. Similarly, a group of people can specialize in making arrows and spears, thus forming a new community of tool-making specialists.

The activity theoretic view implies that one way new communities of practice are created is through increasing specialization and division of labour. An existing community can split into new communities. In this process, the structure of activity and its associated motive structure evolves. Knowledge, in turn, becomes increasingly specialized, new languages emerge, and meaningful interpretations of the world become increasingly varied.

This sociocultural and developmental view assumes that division of labour is part of the natural evolution in cultural development. In this process, beneficial forms of division of labour may become institutionalized through the emergence of specialized

---

[3] Some activity theorists argue that it is important to keep the concepts of object and motive separate. I would argue that one of the strongest points in Leont'ev's activity theory is that it implies that object and motive are actually the same thing. Objects exist as articulations of our motives.

communities of practice. These communities, however, always emerge in the context of existing practice. They therefore have shared 'ancestors' which provide the genetically original foundation for the emerging communities.

An individual inventor may play an important role in this process. New forms of practice can be experimented with and new concepts can be generated to interpret reality in a new way. This experimentation and creation of meaning, however, always occurs using cultural resources that have been produced in history. As Vygotsky put it:

Every inventor, even a genius, is always the outgrowth of his time and environment. His creativity stems from those needs that were created before him, and rests upon those possibilities that, again, exist outside of him . . . No invention or scientific discovery appears before the material and psychological conditions are created that are necessary for its emergence. Creativity is a historically continuous process in which every next form is determined by its preceding ones.   (Vygotsky, 1930; quoted in van der Veer and Valsiner, 1994: p. xi)

Vygotsky also argued that the development of new knowledge occurs through social practice. We don't, for example, first get the idea of hunting and then invent the practice of hunting. Instead, we first hunt, and only after being engaged in the practice of hunting, can we try to construct hunting as a mental phenomenon. Similarly, only when division of labour emerges, can we start to build our identities, languages, and specialized knowledge around the new practice. When we do this, we rely on existing cultural resources and conceptions.

Bijker's discussion on the emergence of Bakelite, therefore, contrasts in an interesting way with the cultural-historical theory of activity. It shows that new technological practices and their associated technological frames do not necessarily emerge only through division of labour and increasing specialization. In the early history of Bakelite, there is no obvious increase in functional differentiation in the social system, or related change in the existing social division of labour. Indeed, it seems that Bakelite is not in any obvious sense determined by its 'preceding forms'. The invention of Bakelite creates material and psychological conditions that are necessary, not for its emergence, but for its existence. As it emerges, new forms of practice, new language and interpretations, and a new conception of synthetic plastics are created. Vygotsky describes a fundamental aspect of innovation and creativity, but the mode of innovation that underlies the emergence of Bakelite relies on a different dynamic. This mode is based on recombining resources produced by existing communities.

7.3 COMBINATORIAL INNOVATION IN AN ECOLOGY OF COMMUNITIES

In the historical description of Bakelite, the locus of innovation is not within a stabilized community of practice. As Bijker notes, the community of Bakelite practice was created only after Baekeland successfully recruited members from several different practices into this new technological frame. Indeed, as Bijker documents, the innovators within the existing plastics community had great difficulty in understanding Baekeland's invention as something that related to plastics.

Bakelite practice was born, instead, by combining existing practices in a new way. Baekeland put together some goals from the plastics technological frame, methods from electrochemical engineering, user communities from electronics and automobile industries; heated them under pressure, and moulded a new industry out of these raw materials.

The Bakelite community of practice, therefore, has a developmentally different origin and dynamic from those communities that emerge through division of labour. Whereas division of labour has its roots in specialization, Baekeland effectively created a new domain for knowledge and practice. As he recruited enough members to this new domain, the domain stabilized and its members became specialists in Bakelite knowledge. The end result is a new community of practice and a new social stock of knowledge that it produces and reproduces. Although in this sense it is similar to those communities that emerge through division of labour, it has developmentally different roots.

The locus of Bakelite innovation cannot be found by studying a specific community. Instead, we have to describe it on the level of an ecology of communities. Such innovation has a combinatorial dynamic. Whereas the evolving division of labour creates new communities as spin-offs from existing practice, combinatorial innovation mixes ingredients from existing communities of practice, and creates a novel domain where a new practice can stabilize.

In the formation of the Bakelite community, four qualitatively different stages can be distinguished. First, someone has to combine existing resources. This agent of innovation we might call the 'combiner'. The combiner operates in a socially peripheral and unstructured domain, appropriating resources for unintended uses. This *ba* of combinatorial invention exists 'in-between' the domains occupied by stabilized forms of practice. As an innovator, the combiner creates an initial form of a new technological opportunity, for example, Bakelite. As a user, the combiner appropriates resources provided by existing communities.

In the second stage, an innovation becomes articulated in relation to social practices. Unintended uses become intended uses. Often the combiner actively tries to promote the innovation for specific uses and recruit user communities. The combiner may or may not be successful in this as the articulation of technology fundamentally depends on the actual user community. There may also be several user communities and the innovation may have several different uses. In the articulation process a network of interests is built around the innovation and the innovation itself becomes defined in relation to this network.

In the third stage, the innovation becomes stabilized and its use may become institutionalized. When the uses are stable enough, a producer community can stabilize. The producer community, therefore, becomes one of the 'users' of the technology in question, and 'the technology' can itself stabilize as a concrete artefact that has existence independent of any specific community.

When technology becomes 'institutionalized' as an interdependence between two or more communities, these communities become symbiotic. In the fourth stage, the innovation couples two or more communities, each of which develops according to its own internal dynamics, but also constrained by its interdependencies.

In the first stage, the combiner is the cognitive centre of the innovation process. However, as he or she operates using a network of human and material collaborators, the effort of the combiner-inventor is mainly about mobilizing these accumulated cognitive and material resources. The combiner is an author in the same sense as a conductor of an orchestra, who uses existing notes and available players to compose a product that can be appropriated and appreciated by the audience.

The product itself, however, is actively created by the audience. For some it may be just noise, and for others just what they wanted and needed. In contrast to music, which is essentially ephemeral as temporal performance, technological products, however, are concrete artefacts. This is of key importance. As soon as a material artefact is generated, it can acquire a life of its own in the field of social interaction. If a concert performance is materialized, for example, by recording it, the audience may get the recording when they leave the concert and keep reproducing the experience without the orchestra. They can listen to a record, for example, to study Bach's theory of invention, or use it to create a suitable atmosphere for a dinner, or for an elevator ride. The recording itself becomes a resource, which the user can appropriate to generate music. At the same time, the control of use shifts to the user and new resources become available that can be applied to make sense of the artefact and combine it in novel ways.

Often a user community does not only use a given technological product, but it consumes them. This continuous flow of interaction binds the communities together. When communities become interdependent, their evolution therefore becomes co-evolution. In the theory of autopoietic systems (Maturana and Varela, 1980) such evolution is called *structural drift*. The communities become *structurally coupled*. The artefact that couples the communities does not need to be defined or problematized any more in any explicit way; instead, it becomes a transparent and routine part of community life. In other words, it becomes a constraint and a resource. In the stable state, the different communities mutually participate in each other's reproduction, and as long as this mutual reproduction goes on, the innovative artefact is stabilized as a nexus of interaction. In a sense, it loses its innovativeness and becomes part of the metabolism of the practice. This is schematically depicted in Fig. 7.1, using Bakelite as an example.

Human societies, however, are different from symbiotic biological systems. In an ecology of communities, structural drift is not the whole story: in a human society resources can be intentionally allocated. For example, Bakelite can be sold to those who pay most for it. Communities can compete about resources. The social ecology of communities and practices, therefore, is complemented by an economy that links and unlinks communities and their resources. A resource allocation economy, therefore, leads to 'creative destruction', as Schumpeter (1975) suggested. The linus themselves, however, are irreducibly social.

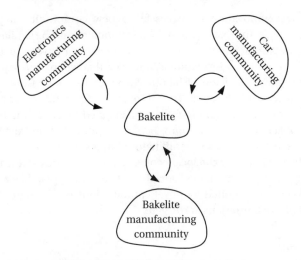

Fig. 7.1. Bakelite as a structural coupler between communities

In a paradoxical way, combinatorial innovation can create true novelty out of old components.[4] In contrast, spin-offs from existing practice always occur from within a given frame. Division of labour and its associated specialization makes the social system more efficient. Evolving division of labour, however, can also create novelty as the emerging practices can drive change in the system of motives and meanings.

There exists therefore a complex social negotiation process where different objects and activities are constructed, institutionalized, changed, and turned into resources. After Bakelite becomes stabilized, it can rapidly be appropriated for purposes of efficient division of labour. Similarly, the Bakelite community itself can become a resource, for example, when it may be enrolled for manufacturing new plastics. The developmental history does not, therefore, determine the role a community will play in the ecology of communities of practice after it becomes stabilized.

The novelty of Bakelite has it source in the fact that its invention used by-products from several practices, which were not trying to produce Bakelite. In a very fundamental sense combinatory innovation is a product of unintended resources. It serendipitously appropriates practices and knowledge generated in other communities and turns them into resources for its own purposes. At the same time—based on its own interests and perspective—it interprets what these resources actually are.

---

[4] This combinatorial character of innovation has often been noted (e.g. Teece, Pisano, and Shuen, 1997; Kogut and Zander, 1997; Nahapiet and Ghoshal, 1998; Hamel, 1999). The actual link between combination and generation of novelty, however, has rarely been explicitly discussed (an important exception being Schon, 1963). In organization theory and knowledge management literature it is often assumed that by putting together old things something new emerges. Epistemologists have always had difficulties with this idea, as logical combinations of existing knowledge, strictly speaking, cannot produce true novelty. Indeed, novelty does not emerge from deterministic combination of existing resources: it can, however, emerge from creative combination of existing resources. As was noted before, this can be observed, for example, by studying J. S. Bach's *Inventions*.

If combinatory innovation is successful, a new community of practice may emerge. In this sense, the end result of the development may be similar to the result of increasing division of labour. An important difference between these two generative modes of community formation is, however, that whereas it is quite simple to intentionally organize division of labour, for example, as divisions and product development teams, it is more difficult to organize for combinatorial innovation. Combinatorial innovation cannot easily be generated by allocating tasks within a given system of activity.[5] In other words, whereas spin-offs from existing communities already have a social basis, combinatorial innovation does not have a social base, or it is located outside the local ecology of communities. Radical innovation, when it occurs through evolution of social practice, may appear like a revolution of the masses. When it is ignited by novel combinations of old ingredients, it may, however, sound like barbarians at the gate.

7.4 LAYERED *ba* AND COMBINATORIAL INNOVATION

We have now described different conceptualizations of knowledge creation in social systems where resource combination and division of labour generate communities, practices, and related social stocks of knowledge. Now we can refine our concepts of innovation spaces.

If we interpret the concept of *ba* in the Kyoto School sense of the term, *ba* is a field of meaning creation where innovative artefacts and their uses become mutually constructed. In such a space, socialization, externalization, articulation, and internalization are difficult to separate. Social learning occurs through interaction, observation, dialogue, and internalization. Externalization occurs, similarly, through interaction, observation, dialogue, and articulation. Dialogue itself can be understood as an interactive process where new meaning is articulated, and where the articulated meaning becomes available for individual and collaborative action. Indeed, the concept of *ba* most naturally seems to apply to a micro-level analysis of meaning creation.

A meso-level description, on the level of social practice, shows that existing technological frames evolve and resources are combined. A macro-level description, in turn, puts communities into an ecology where existing communities of practice are enrolled as members of evolving and emerging frames, where different systems of activity produce resources that are used in other activities, and where some of these activity systems are embedded in organizations and business firms.[6]

In a sense, the four different types of *ba*s proposed by Nonaka and his colleagues, therefore, collapse into two. The systematization *ba* is a place where existing

---

[5] This is one of the reasons why network relations between organizations become prominent in combinatorial innovation.

[6] At that level we therefore have the ecological view on innovation in the sense discussed by Kelly, Kranzberg *et al.* (1978).

resources are combined. The socialization/dialoguing/internalization *ba* is the place where novel resources are created.

In other words, in the *ba* that systemizes knowledge, the focus is on exploitation of knowledge that exists outside the focal knowledge creation space. In the other *ba*s the focus is on creating new knowledge within the focal knowledge creation space through learning, articulation, and externalization.[7]

When the *ba* relies on resources that are embedded in the community that provides the foundation for the *ba*, we can talk about an evolutionary *ba*. When the *ba* relies on resources that are not embedded in the focal community, we can talk about combinatory *ba*. Even in such a combinatory *ba*, resources are produced in systems of activity that have their basis in communities of practice. These resources are always produced in a context of practice, and they are always consumed in a context of practice. The contexts, however, are different.[8]

Although the concept of *ba* and the concept of community of practice have sometimes been viewed as incompatible (e.g. Nonaka, Toyama, and Konno, 2000: 15), they can also be seen as complementary descriptions of the basis of knowledge creation. *Ba*, in the Kyoto School interpretation, is a field of meaning creation, which operates in a cognitive phenomenological domain. Although Nishida's existential phenomenology posited an individual as the focus of his analysis, the individual has to learn and use socially produced cultural resources, such as language and knowledge, to become a cognitive being. Individuals are always social. Community of practice, on the other hand, operates at the level of recurrent collective activity. It defines an interpretation of the world as a socialized thought style, as a specific way it uses tools, language, and knowledge, and as a producer of specific material and immaterial artefacts.

The ongoing meaning processing that occurs within communities of practice can therefore be described using the Kyoto School version of the concept of *ba*. The process of *ba* sustains a specific reality as a meaningful world. When new knowledge and forms of practice are generated in this *ba*, they have to be, however, institutionalized and sedimented into the community thought style and routines. Meaning creation, therefore, is based on a relatively slowly changing community layer, which provides the conventions and concepts that are needed to produce new concepts and conventions. These two layers can be represented as in Fig. 7.2. In a simplified way, we could say that the slowly changing layer of meaning creation

[7] This learning / exploitation distinction, of course, breaks down when we analyse it more carefully. The resources that we exploit have their origin in socialization and learning that has occurred in those communities where we have been members. Indeed, as actor-network theorists (e.g. Law, 1992; Latour, 1999) have argued, an individual is in an important sense a sum of those social networks and tools that he or she can turn into resources in social interaction (cf. Nardi, Whittaker, and Schwartz, 2000). Therefore, all dialogue that produces new knowledge relies on heterogeneity of resources and interpretations that are put into interaction by the different parties in the dialogue. As Bakhtin noted, such dialogue often produces more than it contains.

[8] Therefore combination cannot be understood simply as combination of decontextualized information. Indeed, this is impossible simply because information does not exist without a context (Tuomi, 2000).

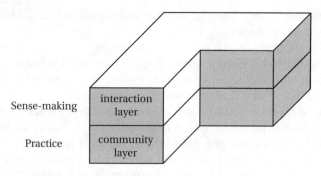

Sense-making   interaction layer

Practice   community layer

Fig. 7.2. A layered *ba*

can be associated with community practices, whereas the rapidly changing and dynamic layer can be associated with communication and sense-making.

The two different levels of analysis, an interaction *ba* and a practice-related community, emerge when we focus on two qualitatively different domains of knowledge creation. When we describe collective production of new technologies and artefacts, the community and the change in its structure of activities and practices become focal. When we describe production of new meaning, the interactions within the *ba* of meaning creation become focal. In the former case, material artefacts and new technologies are created; in the latter case technological frames, concepts, and systems of meaning evolve.

In Fig. 7.2 the lower right quadrant is purposefully missing. In Nonaka's model, this quadrant was associated with the combinatory mode of knowledge creation. As we saw above, a systematization *ba*, however, has a qualitatively different social layer from the originating/dialoguing/exercising *ba*. Innovations can originate in a systematization *ba*, as was seen in the history of Bakelite. These innovations, however, are combinatorial innovations. Their origin is in the interstices of several different communities that have been turned into resources.[9] A schematic description of the structure of the systematization *ba* could look, for example, as in Fig. 7.3.

Innovations, therefore, also have two qualitatively different paths of diffusion. New knowledge can cross the boundary between ongoing interaction and institutionalized social practice, and sediment in the institutions of the focal community.

---

[9] Nonaka's original definition of the lower right quadrant in his SECI-model was based on the idea that explicit knowledge can be combined with explicit knowledge to produce new knowledge. The distinction between explicit and tacit knowledge, however, is problematic as explicit knowledge always requires a background context of tacit knowledge, as was noted before. Instead of explicit knowledge we should probably speak of knowledge resources. In the terms of actor-network theory, explicit knowledge in the 'combination' quadrant is knowledge that has been 'translated'. The concept of resource implies a process of 'black-boxing' which hides much of the complexity of the tacit meaning associated with the knowledge resource. This also means that there are two qualitatively different types of 'explicit' knowledge that are converted in Nonaka's SECI-model. Articulation produces 'externalized' knowledge, whereas combination uses knowledge artefacts that have been translated into resources.

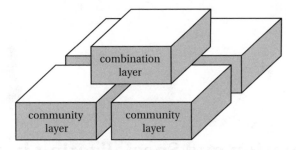

Fig. 7.3. Layers of a systematization *ba*

Innovations, however, can also diffuse by inducing change in external communities. Such induction across communities is often mediated by artefacts. We could simply call the former *vertical diffusion* and the latter *horizontal diffusion*.

Humans are reflective beings, and all material objects have a dual existence as concrete artefacts and as meaningful objects. But, although interpretations and artefacts are closely linked, they also have different dynamics of change. Most importantly, material artefacts, such as new technological products, are always more than what we intend them to be. Therefore, technological artefacts have the capability of surprising us. This, indeed, may be the fundamental reason why in social systems new knowledge is created not only through increasing division of labour, but also by combining products of knowledge in novel and unexpected ways. Sometimes useful combinations result from simple tinkering. Perhaps the fact that some people find it amusing and interesting to break things apart and put them together again shows that humans are genetically innovative beings. Indeed, by looking at the recombination mechanisms in living cells it is easy to see that nature has known the combinatory dynamic of innovation already long before social division of labour became important. The history of the Internet is rich with examples of such tinkering and recombination, and it can be argued that the Internet itself has made the combinatory mode of innovation an increasingly important part of our modern world. It is therefore useful to revisit the history of the Internet to see how combination and evolving specialization interact in the development of technology.

8

Combination and Specialization in the Evolution of the Internet

The different technological frames that provided the basis for ARPANET first became combined when ARPANET was developed and implemented in 1969. ARPANET, as a technological artefact, crystallized ideas that originated from several communities of practice. Its designers often took earlier practices for granted, thus carrying them into technological designs they were creating. For example, it was generally assumed that remote access to time-shared mainframe computers was going to be the main use of the ARPANET. At the same time, the produced artefact created several new communities. At the beginning of the 1970s, ARPANET acted as a focal point that collected several lines of development, concentrated these in time, and rapidly changed itself into a platform on top of which many new innovations were built. Whereas ARPANET in 1969 was an object of innovative activity, in the following years it was turned again and again into a resource for creating other innovations.

The evolution of ARPANET in the 1970s has too many important stages to be covered in detail. I therefore briefly discuss only two developments here. The first is the emergence of email, and the second is the evolution of one central locus of innovation, the Network Working Group.

8.1 EMAIL AS A COMBINATORIAL INNOVATION

Electronic mail has been the main driver for the diffusion of computer networks since the beginning of computer networking (Naughton, 2000: 140–50; Abbate, 1999: 106–11). Electronic mail emerged originally in the context of time-sharing computers, in the early 1960s. These early email systems allowed users of time-shared computers to send messages to other users of the same computer. In 1970 Ray Tomlinson combined an intra-machine email program with his experimental network file copy program, and created a system that could deliver email between two different computers. When the ARPANET project developed a standard for file-transfer, it was soon realized that Tomlinson's program could be used to send email over the ARPANET.

Tomlinson worked for Bolt, Beranek and Newman, the firm that was contracted by ARPA to develop hardware for ARPANET. He had programmed an experimental program to share files between computers and earlier programmed an email program that worked within a single machine. File transfer became a problem and opportunity as the ARPANET developers at BBN connected computers to test the design of ARPANET. There was no explicit project to develop email; instead, Tomlinson used BBN's resources to test whether such an inter-computer email system could be possible.

In 1973 it was estimated that 75 per cent of all network traffic in ARPANET was email. This came as a surprise to many sponsors of the ARPANET. Although email was considered as one possible use of computer networks already in the early phases of the ARPANET project, it was not expected to be important. For example, Larry Roberts did not include electronic mail in the original blueprint for ARPANET. In 1967 he noted that the ability to send messages between users was not an important motivation for a network of scientific computers (Abbate, 1999: 108). As Abbate notes:

In the process of using the network, the ARPANET community developed a new conception of what networking meant. Since the original view of the network planners was that 'resources' meant massive, expensive pieces of hardware or huge databases, they did not anticipate that people would turn out to be the network's most valued resources. Network users challenged the initial assumptions, voting with their packets by sending a huge volume of electronic mail but making relatively little use of remote hardware and software. Through grassroots innovation and thousands of individual choices, the old idea of resource sharing that had propelled the ARPANET project forward was gradually replaced by the idea of the network as a means for bringing people together.   (Abbate, 1999: 111)

Whereas the original goal of the system was to effectively use expensive computer resources, in practice ARPANET was mainly used to send messages across the network. During the 1970s this created continuous tensions between the users of ARPANET and its funders. However, although email was often seen as an illegitimate use of computer resources, it was also seen as a way to expand the use of the network, and therefore accepted.

Tomlinson's email diffused rapidly because it didn't require fundamental change in existing practice. It parsed together existing technologies and provided a new way to accomplish existing actions. Within the ARPANET community, its appropriation required little individual learning and little change in social practice.

In that sense, email was not a major innovation. However, although its appropriation was easy, after it was taken into use it also opened new possibilities. The rapid expansion of email use, therefore, can to a large extent be attributed to the fact that it was introduced as a tool for an old and well-known practice; yet, it also had the capability to transform this practice into multiple new forms. As a result, the take-off for email was fast, and it kept expanding far beyond its original scope. *After* email was in wide use, new systems and practices, such as mailing lists, newsgroups, and collaboration systems could be introduced. But if email had been introduced *as* a mailing list system, for example, it is quite conceivable that it would never have taken off.

Email was an explosive innovation. It ignited easily and generated increasing amounts of energy as new users invented new uses for it. This rapid diffusion of email, however, did not result from a simple creation of a 'critical mass' or 'network effects'. Instead, there were two quite different mechanisms at work. Simple 'diffusion' occurred as existing practices substituted electronic media for previous technologies. The new technological opportunity of email was simply taken into use. As soon as this happened, however, a new expansive dynamic set in and email started to acquire new meanings.

In Bakhtin's terms, Tomlinson utilized the existing genres of computer messaging and file transfer, at the same time creating the infrastructure for new genres. In activity theoretic terms, email was introduced at the lowest level of the activity hierarchy. It entered the system of activity at the level of operations, as a tool for existing actions. At the same time, it made many new combinations of actions effective, thus reorganizing the system of activity and its motive structure. In this process, many communities of practice were created.

Networked email was a relatively simple incremental modification of existing technology. Its results, however, were radical. It transformed the way people understood computer networking in general, and ARPANET in particular. Its very rapid diffusion indicates that it addressed a very generic need, without much limiting the way it was used. In this sense, we could say that it added a new medium on top of a medium that was itself just emerging: the ARPANET. Packet-switching networks and email mutually constructed each other.

At the same time, the conditions for social interaction changed; first, on a laboratory scale and then, more broadly. Email converted the ARPANET into a system of computer-mediated communication. As a result, new communities of practice started to emerge, which extensively used this new medium. Computer networks became the infrastructure for social processes and virtual communities became possible. Today, for example, there are over 40,000 Usenet newsgroups on the Internet. Internet communities now range from discursive communities that produce and reproduce identities to technology development communities, which generate new technological artefacts. The Linux development model is in many ways a typical technology development community on the Internet and I will analyse it in detail in a later chapter.

Implementation of new combinations often occurs very rapidly in software technology. Software is qualitatively different from other technologies as code is both representation and implementation. Whereas in other technologies a description is never the artefact that is described, in software functionality and its description are embedded into the same piece of code. This makes software uniquely mobile. The fundamental reason for the mobility of software is not—as is often assumed—that software code can be transmitted in a digital form. In other areas of technology, knowledge has a large tacit component, as for example Collins (1975; 1987) has shown. The mobility of software, and the possibility of appropriating it for unintended uses, depends on the fact that there is no distinction between semantics and syntax in software.[1]

---

[1] There are, of course, large amounts contextual knowledge that is needed to make sense of software or to use it (cf. Tuomi, 2000).

It is also rather remarkable that due to the combinatorial technical nature of email, it had no well-defined developer community. Software can be written so that its recombination and reuse is easy. It is often possible for a single individual to realize new combinations independent of any specific stabilized community of practice. In other words, software quickly creates virtual worlds but can also make them real. In the history of email, the developer community was made redundant by fixing and standardizing the basic functionality.

Without a stable developer community, the stability of the emerging technological frame to a large extent depends on practices of use. As we now know, the introduction of email rapidly led to new uses and practices, which in turn generated new complementary innovations, such as email readers, email gateways, email list servers, news servers, and domain name standards. In contrast to many historical regimes of innovation, the locus of software-based innovation is therefore often located in communities of use.

A particularly important community, of course, is the community that maintains and operates those services that are required to keep email moving. These people are often called system operators. Although they don't necessarily develop the basic functionality of email, and indeed sometimes hate the idea that someone would like to do it, their practice includes the removing of bottlenecks in email traffic, the development of configuration scripts, and the debugging of problems in email delivery. The operator community is inherently conservative as its members are usually paid to maintain things as they are. The locus of innovation in email, therefore, is distributed across downstream communities.

Email is an exceptional combinatorial innovation. Tomlinson was able to patch it together without enrolling any community or without creating a new community from scratch. He had personally developed both single computer email and networked file transfer programs and he only had to convince himself to build a working email system. The cultural and cognitive gap between the two pieces of software was minimal. Not much learning and no social learning was needed to make the program components work together.

Furthermore, the early ARPANET community easily played the dual role of using and managing the email system. Its use was trivial as it extended the existing practice of sending conventional mail. The user community had no difficulties in managing and operating the system as the users were specialists in computer networking. In effect, the different communities that generated resources, combined them, and used and maintained the resulting system collapsed into one.

This is not always the case. Napster, for example, relies on file transfer and the Internet domain name system, making it similar to email (Shirky, 2001). As playing music is an existing and generic activity, the use of Napster could spread easily. Music listeners and Napster developers, however, do not necessarily belong to the same communities of practice. Moreover, the resources used in Napster are produced by communities that don't have much overlap with computer software developers. To invent Napster, one also has to invent new revenue models for the recording industry. Otherwise, the lawyers come knocking at the door.

8.2 ARPANET ECOLOGY AND THE EVOLUTION OF THE NETWORK WORKING GROUP

In the history of the Internet, a critical role was played by the Network Working Group (NWG). It was formed early in the ARPANET project to discuss potential uses of the network, and to create specifications for programs that interacted through the network. It was to a large extent a self-organized group of graduate students. The group had no visible authority or decision-making power. Yet it successfully developed the host-to-host program that became the core software for ARPANET. During subsequent years the NWG evolved into an ecology of communities that created most of the ARPANET and Internet technology and applications. In the history of the NWG, we can in very concrete terms see how communities emerge, disappear, and transform themselves.

An embryonic form of the Network Working Group can be found in a group of Principal Investigators of ARPA/IPTO-funded research sites. When IPTO was planning to launch the ARPANET project, it used the Principal Investigators' meeting in April of 1967 to get feedback on the idea of building a computer network. The idea of computer networking was not greeted with great enthusiasm as some Principal Investigators saw the network as an alternative to buying them more computers (Abbate, 1999: 50). Yet, several important contributions were made in the conference and as a result of it. At the meeting it was decided that there had to be agreement on conventions for transmitting characters and binary blocks, transmission error checking, retransmission, and computer and user identification. Frank Westervelt was chosen to write a proposal on these topics and a communication group was formed to study the related questions (Hauben and Hauben, 1997: ch 6).

One of the Principal Investigators, Wesley Clark of Washington University at St Louis, proposed an architecture that separated the core communication network from the mainframe computers that used the network. Instead of directly connecting mainframes—also known as the 'hosts'—to each other, each host was connected to an 'Interface Message Processor', or IMP. The IMPs, in turn, were connected to each other. This 'layered' architecture meant that each different mainframe had to be able to communicate only with an IMP—a program on a relatively cheap minicomputer—and not with all the different mainframes connected to the network. This greatly reduced the complexity of the network programming, and made it possible to develop the core network independently from the mainframe host applications that used the network.

These basic ideas were discussed and presented at the ACM symposium on 1–3 October 1967 (Naughton, 2000: 129–31). In the same symposium Roger Scantlebury from the NPL described the packet-switching network being developed in the UK. After the session, a number of attendees gathered to discuss network design, and Scantlebury and his colleagues advocated packet-switching as a way to implement the network planned by ARPA (Abbate, 1999: 38; Hafner and Lyon, 1998: 76). These ideas were further discussed at a meeting organized by ARPA in October. Elmer Shapiro of SRI was given the task of writing a report on ARPANET architecture, and

COMBINATION AND SPECIALIZATION 143

Larry Roberts and Barry Wessler of ARPA wrote the specification for the IMPs based on Shapiro's work.[2] This specification was discussed in the Principal Investigator meeting in June 1968, and a program plan for 'Resource Sharing Computer Networks' was submitted to the ARPA Director on 3 June 1968. Based on the accepted program plan, a competitive Request for Quotation was mailed to 140 potential IMP developers. ARPA received twelve proposals, narrowed them to four, and granted the contract finally to BBN (Hauben and Hauben, 1997: ch. 6).

The IMPs were necessary to build the communication network that mainframes were to use for their network applications. The applications, however, had to be defined. In 1968 it was unclear what those applications would be. Elmer Shapiro was therefore asked to organize a group that would define applications that the network would support. This group, which became known as the Network Working Group, originally consisted of representatives of the first ARPANET sites. The completion report of the ARPANET project later noted that at the beginning it was not clear what the group was supposed to do:

To provide the hosts with a little impetus to work on the host-to-host problems, ARPA assigned Elmer Shapiro of SRI 'to make something happen', a typically vague ARPA assignment. Shapiro called a meeting in the summer of 1968 which was attended by programmers from several of the first hosts to be connected to the network. Individuals who were present have said that it was clear from the meeting at the time, no one had even any clear notions of what the fundamental host-to-host issues might be. (ARPANET Completion report draft III, quoted in Hauben and Hauben, 1997: ch. 6)

The meeting was chaired by Elmer Shapiro and other attendees included Jeff Ruflinson of SRI, Ron Stoughton from UCSB, Steve Carr from University of Utah, and Steve Crocker from UCLA.[3] Shapiro opened the meeting with a list of questions on how the IMPs and hosts would be connected, what hosts would say to each other, and what applications would be supported.

No one had any answers, but the prospects seemed exciting. We found ourselves imagining all kinds of possibilities—interactive graphics, cooperating processes, automatic data base query, electronic mail—but no one knew where to begin. We weren't sure whether there was really room to think hard about these problems; surely someone from the east would be along by and by to bring the word. But we did come to one conclusion: We ought to meet again. Over the next several months, we managed to parlay that idea into a series of exchange meetings at each of our sites, thereby setting the most important precedent in protocol design. (Crocker, 1987)

According to Crocker, the first few meetings were quite tenuous. The group was informal without official charter:

Most of us were graduate students and we expected that a professional crew would show up eventually to take over the problems we were dealing with. Without clear definition of what

---

[2] The SRI report was issued in Dec. 1968 as 'A Study of Computer Network Design Parameters'.
[3] These are the attendees as recalled by Steve Crocker. He did, however, note that records of this meeting are lost and there might have been other attendees (Crocker, 1987).

the host-IMP interface would look like, or even what functions the IMP would provide, we focused on exotic ideas. (Crocker, 1987)

The initial group met a handful of times in the summer and autumn of 1968 and winter of 1969, discussing mainly conceptual ideas. One of these was a language that could be used to download small interpretative programs at the beginning of a network session, and which could then control the interactions across the network (Crocker, 1999).[4]

The NWG work moved from concepts to the concrete problem of utilizing the underlying IMP network when BBN issued its host-IMP specification in the spring of 1969. Around that time, the NWG decided that it was necessary to write down notes that documented the meetings. Steve Crocker, who at that time had become the chair of the NWG, started to put together memos about the topics discussed in the NWG meetings. These memos were called 'Request for Comments'. These RFCs later became a major form of developing ARPANET and Internet technology.

8.2.1 RFCs and the Four Waves of Internet Development

According to Crocker, the early history of RFCs had a definitive impact on the evolution of the Internet and the way its development eventually became organized:

Two all-important aspects of the early work deserve mention, although they're completely evident to anyone who participates in the process today. First, the technical direction we chose from the beginning was an open architecture based on multiple layers of protocol. We were frankly too scared to imagine that we could define an all-inclusive set of protocols that would serve indefinitely. We envisioned a continual process of evolution and addition, and obviously this is what's happened.

The RFCs themselves also represented a certain sense of fear. After several months of meetings, we felt obliged to write down our thoughts. We parceled out the work and wrote the initial batch of memos. In addition to participating in the technical design, I took on the administrative function of setting up a simple scheme for numbering and distributing the notes. Mindful that our group was informal, junior and unchartered, I wanted to emphasize these notes were the beginning of a dialog and not an assertion of control. (Crocker, 1999)

The first two RFCs discussed the requirements of the host software. The third RFC described the documentation conventions to be used in the RFCs. It pointed out that the NWG was concerned with issues related to the host software, the strategies for using the network, and the initial experiments with the network. The efforts of the NWG were to be reported through the RFC notes. This early document, written in April 1969, set the tone for later ARPANET and Internet development:

The content of a NWG note may be any thought, suggestion, etc. related to the Host software or other aspect of the network. Notes are encouraged to be timely rather than polished. Philosophical positions without examples or other specifics, specific suggestions or

---

[4] This idea was actually implemented in the Java language and ActiveX, about three decades later.

implementation techniques without introductory or background explanation, and explicit questions without any attempted answers are all acceptable. The minimum length for a NWG note is one sentence.   (Crocker, 1969)

In commenting on this third RFC, Jake Feinler notes:

Thus by the time the third RFC was published, many of the concepts of how to do business in this new networking environment had been established—there would be a working group of implementers (NWG) actually discussing and trying things out; ideas were to be free-wheeling; communications would be informal; documents would be deposited (online when possible) at the NIC and distributed freely to members of the working group; and anyone with something to contribute could come to the party. With this one document a swath was instantly cut through miles of red tape and pedantic process. Was this radical for the times or what! And we were only up to RFC 3!   (Feinler, 1999)

The number of documents published in the RFC series shows that there have been three or four waves of Internet-technology. This can be seen in Fig. 8.1. The first wave peaked around 1972. This is the time when key ARPANET applications, such as email and the FTP file transfer protocol, were designed. As the use of ARPANET grew and the original development goals were achieved, ARPA started to look for an organization that could take care of the operational management of the ARPANET and provide nationwide public service (Abbate, 1999: 134–5). After unsuccessful attempts to find a commercial operator, ARPANET was temporarily transferred to the Defense Communications Agency (DCA) in 1975. ARPA itself focused its efforts on the problem of networking multiple networks. The first internetwork experiments were conducted in 1975 (Abbate, 1999: 130–1).

After the DCA started to operate the ARPANET, its future was uncertain for several years. ARPANET was built to develop computer networking technology. The experiment was successful, but ARPANET itself was still clearly an experimental system. The Defense Department was developing its own computer networks,

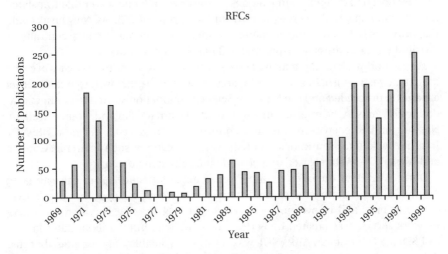

Fig. 8.1. The waves of Internet technology development

and planned to replace ARPANET after these military-grade operational networks were ready. First the DCA considered dismantling the ARPANET, but eventually the agency concluded that there was also a role for a research-oriented network (Abbate, 1999: 139).

The parallel use of ARPANET and the operational military networks, however, made it clear that internetworking was going to be a problem also in the future. In 1980 the Office of the Secretary of Defense adopted the new Internet protocols as official standards. The existing ARPANET sites were encouraged to start developing software for the new TCP/IP Internet protocols. In March of 1981, DCA's ARPANET manager announced that all ARPANET hosts would be required to replace the old ARPANET host-protocol with the new TCP/IP by January 1983.

The new Internet protocols, therefore, became the centre of activity for the ARPANET community. An increasing number of Internet-related notes started to appear in the RFC series, and in 1983 the Internet Experimentation Notes were merged with the original RFC series. In that same year, also a new system for handling host names was introduced. The original ARPANET host name system required that each computer kept a full list of all host names and their corresponding ARPANET addresses in a file. When the number of hosts increased, this became a major problem. The new domain name system divided the Internet into separate domains, each managed by a host that managed host names in its own domain. Instead of maintaining all host names in a local file, a host could send a request to the domain name server and retrieve the Internet address of the host with a given name. Six top-level domains were selected to separate educational, military, governmental, commercial, non-commercial organizations, and network resources. These top-level domains were distinguished by adding .edu, .mil, .gov, .com, .org, and .net to their names, respectively. Within each top-level domain the name system was divided into domains. Each domain had a host that provided information on the addresses of the host within the domain. To communicate with a host, another host had to send a name request to the domain name server that translated the Internet name to the corresponding Internet protocol address. New hosts could therefore easily be added to domains without coordinating changes centrally or distributing the address information to all hosts in the network.

The second peak in RFC activity occured around 1983. After that, clear waves are difficult to distinguish. In 1987 the RFC activity started to grow as the global Internet expansion proceeded and the National Science Foundation started to connect academic networks to the Internet. In 1988 Canada, Denmark, Finland, France, Iceland, Norway, and Sweden were the first countries to connect to the new NSFNET Internet backbone. New information-sharing applications, such as Archie, Gopher, and World Wide Web released in 1990 and 1991, gained in popularity.

During that time, the Internet underwent a major transformation. The estimated number of hosts connected to the Internet went from over 1,000 in 1984, to over 10,000 in 1987, and to over 100,000 in 1989. In 1989 the first commercial TCP/IP network service was created by Performance Systems International (later known as PSINet). The original ARPANET was in theory available only for sites that had defence contracts. NSF expanded the network to all academic institutions, and also linked its network to networks outside the US. The use of the NSFNET Internet backbone, however, was still restricted. According to the NSF's Acceptable Use Policy, the backbone was reserved for open research and education, and commercial use was not permitted. As a result, in the early 1990s several commercial service providers started to launch their own TCP/IP services. As it became increasingly obvious that there would be alternative commercial Internet services, NSF decided in 1991 that its academic backbone services could be outsourced. The privatization of the Internet created a new peak of activity around 1993.

In 1993 the Internet also broke through in the public media. The World Wide Web and Mosaic created front-page publicity for the Internet. The Internet moved from the academic world first to business use and quickly became a commodity that is integrated with all PCs. This created a new entrepreneurial wave of Internet development, and Internet standards became important for many new players. Judging from Fig. 8.1, the peak of this activity seemed to occur around 1998.

8.2.2 Organization and Community Ecology

The ARPANET architecture thus led to a natural division of labour: someone had to develop the hardware and software for the core network, and someone else had to develop computer programs that used the network. The NWG was set up for the latter task. The former task was given to Bolt, Beranek and Newman, after a competitive bid in 1968. A Network Information Center (NIC) was also set up to manage information related to ARPANET development. This task was contracted to SRI's Augmentation Research Center. When ARPANET expanded beyond its first four nodes in the beginning of 1970, it became apparent that the operation of the network required support. ARPA therefore contracted BBN to set up a Network Control Center (NCC), which monitored the network and gave support to network users.

As the network expanded, the NWG grew rapidly from half a dozen participants to about one hundred, in a couple of years. Soon several informal working groups and special interest groups were formed within NWG. These groups were set up around interest areas and specific program development activities, and often they were transitory.

ARPANET activity expanded rapidly in 1972 when ARPA directors decided that the system should be demonstrated at the First International Conference of Computer Communications. Several applications were developed, and the possibilities of computer networking were successfully demonstrated to an audience of about a thousand experts. During the conference, an International Network Working Group (INWG) was formed. INWG brought together people who were developing packet-switching networks in different countries, including the US ARPANET, the UK National Physical Laboratory's newly designed Mark II, and the French Cyclades. The INWG was chaired by one core member of the ARPANET NWG, Vinton Cerf. The INWG reorganized itself a year later to be a part of the International Federation for Information Processing (IFIP). As a result the INWG transformed itself into IFIP Working Group 6.1.

As the ARPANET grew, the number of users soon bypassed the number of ARPANET developers. In 1973 there was an attempt to set up a group to represent users, but when it tried to get involved with the planning of ARPANET architecture, ARPA ended funding for the group.

The various international network projects made it increasingly visible that ARPANET was going to be only one network among several different networks. ARPA was also funding packet-switching networks that used radio and satellite connections. As a result, it became clear that some mechanisms should be developed to interconnect the various networks. ARPA started to fund research on interconnected networks in 1973, and soon this became the main focus for ARPA's network activities. Vinton Cerf and Robert Kahn, at that time head of ARPA/IPTO, wrote the first paper on Internet design in the summer of 1973. Cerf continued this work under a contract from ARPA, and became a program manager for ARPA's network projects in 1976.

To develop the Internet, Cerf set up a group, along the lines of the original NWG, which he called the Internet Working Group (IWG). When the size of this group grew, Cerf reorganized it in 1979 into three groups. One of these, the International Cooperation Board (ICB) focused on coordinating European network projects and ARPA's Internet Program; another was an inclusive Internet Research Group (IRG), which provided a platform of open discussions on internetting; and the third group was the Internet Configuration Control Board, which became a decision-making authority in the Internet Program. Although the original NWG still existed in the form of informal working groups, after ARPANET switched to Internet technology in 1983, the original NWG-related groups effectively became part of the Internet organization.

The organizational structure continued its evolution, reflecting emerging new needs and the growth of the system. In 1983 the ICCB was replaced by the Internet Activities Board (IAB) and a set of Task Forces were created under it. The IAB consisted of the chairs of the Task Forces. In this transition, the earlier ICCB group effectively renamed itself as IAB: each ICCB member was given his own Task Force and the original ICCB members became members of the new IAB. This evolution of the ARPANET development structure is shown in Fig. 8.2.

8.2.3 Spin-off and Resource Combination in the NWG

By looking at the somewhat simplified Fig. 8.2, several characteristics of community development become visible. First, the ARPANET project explicitly created groups based on division of labour. When these groups started to work on their tasks, they quickly evolved into communities of practice. Each community had its own practices, goals, values, and tools. Due to their common origin many of the communities, however, also shared values, practices, and tools. An important common practice was the use of RFCs, and the ARPANET itself became a common tool to distribute these RFCs. This practice was imitated in many communities that emerged

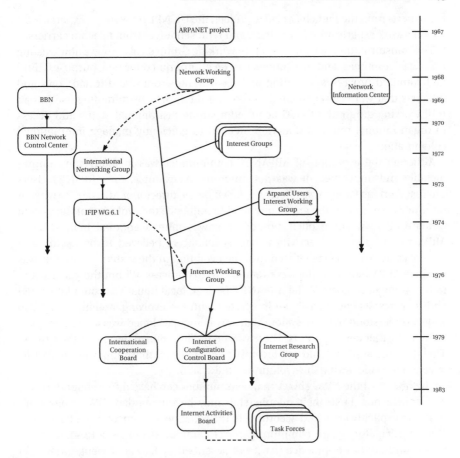

Fig. 8.2. The division of labour in the early phases of Internet development

from the NWG, including the Internet Working Group. The IWG published its own Internet Experiment Notes series until it was merged into the RFC series in 1983 (Cerf, 1999).

When ARPANET became a working system, The Network Working Group expanded together with the use of the network. After programs such as remote access (telnet), file transfer (ftp), and email became available, specialized development communities became increasingly independent of location. The new communities were to a large extent virtual. They emerged in the context of NWG, and later in an ecology of NWG-related communities. These communities developed technologies that improved the network as an infrastructure, thus creating tools for each other.

Communities, however, did not always emerge simply as a result of specialization. The INWG, for example, was a community that brought together several existing communities. Through the INWG, the NWG recruited resources that it could not have developed itself. The membership of INWG included, for example, key experts from the French Cyclades project, the UK NPL networking experts, local area network experts from Xerox Parc, and representatives from telecom carriers.

The transformation of INWG into IFIP Working Group 6.1 also shows how existing practices, resources, and institutions can be appropriated for new purposes. IFIP[5] had standard processes of setting up new working groups, and it had goals and interests that were closely related to those of the INWG. By reframing itself as an IFIP working group, the INWG acquired a strong institutional status and access to organizational routines that were needed to efficiently manage international collaboration.

Although the evolution of ARPANET communities was in many ways unpredictable and unplanned, it was not random. For example, the ARPANET Users Interest Working Group was effectively killed by the director of ARPA/IPTO when it became apparent that the group intended to influence the evolution of the system (Abbate, 1999: 95). This explicit show of power was quite exceptional. In most cases, ARPA simply promoted activities that its managers believed to be useful, and there was no need to explicitly control the evolution of the system. Partly this was because ARPA had no serious contenders in this process. Within the generic and rather ambiguous goals of the ARPANET project, local initiatives could be started and their results could easily be integrated with the evolving system. It was also widely understood that the system that was being developed was a complex platform for further developments. There was no predetermined design for the system. Therefore the system could opportunistically utilize all those innovations that its developers made during the evolution of the system.

The history of the NWG shows that communities can have diffuse boundaries in two dimensions. First, their membership may be unbounded. NWG communities were explicitly launched according to the principle that anyone can be a legitimate participator in the community. NWG communities, therefore, differ from those communities of practice that were described by Lave and Wenger (1991). In Internet communities, the membership is 'graded': some members are in the core of the community and others may be more loosely associated with a given community. Inclusion, however, may smoothly move from non-existent to tight. As long as the peripheral participants are simply observing the activities of the more central participants, peripheral participants are more or less invisible in the community. Only when they become increasingly active, can the impact of their presence be detected. In contrast to traditional communities, Internet communities are therefore different: in economic terms, peripheral participation is costless to the community. This change in the dynamics of communities is fundamentally created by technology.

Communities also have a diffuse membership across time. Participators do not deterministically move from periphery to centre; instead, some participators just

---

[5] IFIP, or International Federation for Information Processing, is an international and apolitical organization that coordinates research and development in the information technology area. Its Technical Committee 6 (TC 6) deals with communication systems, and IFIP Working Group 6.1 is 'Architectures and Protocols for Distributed Systems'. Today, there are several Internet-related working groups organized under TC 6.

visit the periphery, while others rapidly move to the centre, just to disappear at some subsequent time. Due to the informal and voluntary nature of many Internet development communities, the centrality in a community is to a large extent determined by the track record and activity of the members. Centrality therefore reflects the shared memory of the community. The dynamics of NWG communities differ from many earlier communities because their memory is recorded in great detail in a medium that can be used to recreate the history of the community in question.

As NWG communities have exceptionally good organizational memory, one might expect that their technological frames would become rigid. With almost total recall, it is difficult to unlearn and find new ways to do things (Tuomi, 1996). The history of the evolution of ARPANET and the Internet, however, shows that this has not happened: many central values and goals of the developers have changed quite radically during the years.

One important reason for this seems to be that the original technological frame is constantly tested in practice. A current technological frame guides the development, but the results of the development often break the frame. For example, when

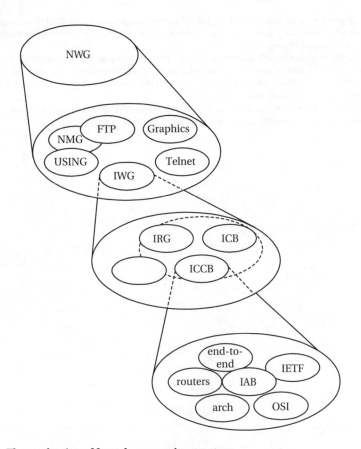

Fig. 8.3. The production of fractal community structure

the Internet became interesting enough to attract commercial developers and users, new organizational structures were set up to make commercial use possible. When it became obvious that the goals of the US ARPA and Defense Department were becoming increasingly difficult to match with commercial interests and the requirements for global networking, the descendants of NWG finally formed the Internet Society, in 1992. As it became clear during the 1990s that the Internet was becoming a major driver in socio-economic development, the Internet Society created in 1999 the Internet Societal Task Force, moving the original focus of NWG from technological issues increasingly to issues related to social development.[6]

The evolution of the NWG communities shows that communities have a 'fractal' character in the time dimension.[7] The evolving division of labour creates communities, which further spin off new communities.[8] As communities do not have well-defined boundaries in time, one can 'zoom in' to a specific community and find its origins in the developmentally earlier forms of community structure. This 'fractalization process' is depicted in Fig. 8.3, which schematically shows the evolution of the NWG during the first fifteen years of Internet development.

[6] The Internet Societal Task Force was closed down in Dec. 2001 as it had difficulties in producing concrete results of its work. The discussion continues at the isdf (internet societal discussion forum) mailing list, managed by majordomo@www.isoc.org.

[7] The fractal model of organization was discussed in Tuomi (1999: 261–75), where it was proposed that organizations could be viewed as 'communities within communities'. The developmental history of such organizational structures shows that part of this 'fractal structure' is generated as a result of ongoing specialization and division of labour.

[8] This developmental view on communities of practice could be understood as a basis of implementing the cellular organization form discussed by Miles *et al.* (1997).

9

Retrospection and Attribution in the History of Arpanet and the Internet

The traditional linear model of innovation implies that the source of innovation lies upstream. Causal chains and 'streams', however, can be defined only after a specific interpretation of an innovation is selected. What we find from 'upstream' depends on how we map the terrain. This is theoretically important when we analyse the loci of innovation, but it also has important practical implications, for example, for attribution of intellectual property rights. When we discuss the actors and agents of innovation, their importance depends on historical reconstruction.

Authorship and reputation are always assigned retrospectively. At the beginning of the twentieth century, it was estimated that there were about 1,000 paintings by Rembrandt. Since then, the number has been cut down to 700, then to 630, and to 420. Ongoing research by the Rembrandt Research Project in Amsterdam will cut the number further (Alpers, 1988). Many of the well-known 'Rembrandt' paintings were not painted by Rembrandt. As Alpers describes, often they were works of Rembrandt's students and by artists influenced by his style.

Indeed, in Rembrandt's studio the concept of authorship acquires a historically new meaning and becomes an integral part of the economy of the art market. The price of a Rembrandt painting, in turn, depends on our theory of authorship and agency. Markets adopt a specific theory of authorship, and make authenticity a question of the identity of the painter, even when it may be impossible to define exactly who the painter is, or when the end result is a product of collaborative work. As Alpers (1988: 143) notes, Rembrandt willingly signed works painted by others. Moreover, he got other artists to pass themselves off as him. In other times and other cultures, the authorship for Rembrandt's works could have been claimed by his patrons. In another culture, the authorship could have been claimed by the maker of the pigments that Rembrandt used for his paints. Historically, this problem becomes an acute problem in the case of Rembrandt, as he is one of the first artists to emphasize the role of an individual author. As Alpers puts it:

The case of Rembrandt makes it particularly clear that the question of attribution is not the same as the question of originality or of invention. Where, then, does 'quality' lie? That individuality claimed by Rembrandt's mode of painting, an individuality which, however, is produced in a workshop situation, presents the problem in a particularly complex way. (Alpers, 1988: 125)

History has a very selective memory. Many artists who were well known and highly respected authors in their own time have rapidly become forgotten. The retrospective prominence of authors has little to do with the quality of their work. As Gladys and Kurt Lang (1988) have shown, contemporary peer recognition does not predict the survival of artistic reputation. Even during the lifetime of an author, the credit often goes to someone else. Robert Merton (1973: 439–59; 1988; 1995) introduced the concept of the 'Matthew Effect' to describe the allocation of credit among authors of multiple discoveries or collaborators, arguing that the 'rich are likely to get richer'. Stephen Cole (1970) and Harriet Zuckerman (1988) showed that those scientists who are located to central positions in the social system of science tend to get most visibility and credit. And, as common sense might hint, those who have accumulated strong reputations have a strong voice when history is told (Gamson, 1966).

The problem of allocating authorship is clearly seen in the phenomenon of eponymy. Eponymy associates a specific idea, phenomenon, or result with a person, as in Gaussian distribution, Planck's constant, Halley's comet, Rorschach test, or Tobin tax. Based on his studies on the history of statistics, Stephen Stigler (1999: 277) proposed his own 'Stigler's Law of Eponymy'. In its simplest and strongest form it says: 'No scientific discovery is named after its original discoverer'.

History is important because it underlies our concept of progress. When innovation is seen as progress, we implicitly generate the linear model of innovation. Innovation becomes a solution to a problem, and we assume that the problem existed before the innovation, waiting to be solved. We further assume that the inventor becomes an agent of action and an author of its consequences. This theory is an integral part of modern concepts of authorship, ownership, and economy. When historical accounts of technological development fit this model well, we think the stories are accurate. When they reveal the complexity of social interaction and development, they remain confusing and without a clear storyline.

Mircea Eliade (1991) argued that in traditional cultures events become meaningful to the extent that they repeat mythical archetypes. The collective mechanisms of attribution and interpretation are in this sense 'pre-modern'. Innovation and innovators become meaningful when they fit the mythical models of innovation. As Eliade showed, recall of meaningful stories quickly fills in the missing details and puts the actors in their expected roles. Halbwachs (1980; 1992), Douglas (1987), and others have noted the same point. Prototypical stories are well recalled and often reproduced.

Histories are sources of meaning and they are actively organized to tell a story for specific purposes (MacIntyre, 1981; Czarniawska, 1997). Innovations, therefore, are not only reconstructed as historical facts. History is always told in a current context and it is produced at the same time as it is recalled. This hits the core of the traditional model of innovation.

9.1 'THE FIRST PAPER ON PACKET-SWITCHING THEORY'

One example of this process can be found in some of the most authoritative accounts of the history of ARPANET and the Internet. In his recent interviews and papers, Larry Roberts claims that Leonard Kleinrock's work at MIT was crucial in helping Roberts to realize that packet-switched networks were the way to go. The role of MIT becomes emphasized in this story, possibly because Kleinrock did his Ph.D. at the same research laboratory with Roberts and, perhaps, because many students of Kleinrock later become important contributors in the development of the Internet. Also in the 'Brief History of the Internet', written by Roberts, Kleinrock and others (Leiner, Cerf, *et al.*, 2000) and published through the Internet Society, Leonard Kleinrock's work is mentioned as the primary source of convincing Roberts that packet-switching would be possible. This rather authoritative version of the history of the Internet also notes that Kleinrock published the first paper on packet-switching theory in July 1961.

It is, however, not obvious that Kleinrock's paper (Kleinrock, 1961) should be considered as the first paper in packet-switching theory. The paper is a Ph.D. proposal, which describes statistical methods that had been used for modelling telephone networks since the early decades of the twentieth century, as the paper notes. The proposal mentions computers twice, in the context of simulating traffic in the store-and-forward networks using a computer. Without the benefit of hindsight, and Baran's and Davies's work on packet-switching, it would probably be difficult to understand Kleinrock's paper as 'the first paper on packet-switching theory'. The paper itself is a well-written proposal, but indicates the difficulty of reading history without seeing it through later developments.

The impression given by the 'Brief History of the Internet' is not historically very accurate but it fits well with the traditional model of innovation as a logical progress of improved technology. According to this account Kleinrock developed the theory for packet-switched networks, Roberts went on to test it by building the first wide-area network between MIT and SDC with Thomas Merrill, and the result of this experiment was the realization that packet-switching was required as Kleinrock had predicted. The logic of events is clear and history unfolds in an organized and linear way. At the same time, the key actors become well defined.

This emphasis on a logical progression of ideas reflects the traditional linear model of innovation. Without well-defined heroes of innovation there cannot be heroic innovation. This requires, however, that the meaning of events and their timing have to be adjusted. This version of the history of the Internet, for example, notes that: 'It happened that the work at MIT (1961–1967), at RAND (1962–1965), and at NPL (1964–1967) had all proceeded in parallel without any of the researchers knowing about the other work'. It is, however, difficult to match the years mentioned to any available documents, and it is well known that there were many interactions and interdependencies between RAND, NPL, and IPTO.[1]

---

[1] Roberts has presented this view of the history of the Internet, for example, in a recent presentation at SIGCOMM 99 (Roberts, 1999*b*).

In his 'Internet Chronology', Roberts (1999*a*) notes that 'the ARPANET and Internet stemmed from the MIT work of Licklider, Kleinrock and Roberts, and had no relation to Baran's work'. In discussing the impact of Baran's work, Roberts writes: 'Roberts read the Rand work and met with Baran . . . the Rand work had no significant impact on the ARPANET plans and Internet history.'

Earlier, in O'Neill (1995: 78), Roberts described his first encounter with Baran's work, in October 1967, as a kind of revelation: 'I got this huge collection of reports back at the office, which were sitting around the ARPA office, and suddenly I learned how to route packets. So we talked to Paul and used all of his concepts and put together the proposal . . . '. In June 1968 Roberts described the ARPANET as a demonstration of the distributed network recommended in the RAND study.

The independence of MIT work on research at RAND is also somewhat unclear. For example, Kleinrock used Baran as a reference in his Ph.D. work in 1963 (Abbate, 1999: 225). The first public presentation of the ARPANET concept by Roberts in October 1967, at the ACM Conference in Gatlinburg, is still based on a store-and-forward message network, without well-defined packets. It was during this conference that Roberts first learned about Davies's and Baran's work on packet-switching.

The Internet is a major innovation with huge social consequences. The way society allocates credit for this innovation has consequences, and an 'unfair' allocation of credit quickly creates controversy. For example, when the US Vice President Al Gore was interviewed on CNN in 1999, he was interpreted as claiming that he invented the Internet. The sentence which generated a lot of comment was: 'During my service in the United States Congress, I took the initiative in creating the Internet' (McCullagh, 1999; 2000).

Many commentators rushed to note that this claim was absurd. But, as Quarterman (1999) points out, it is not necessarily as absurd as many believe. Senator Gore was holding hearings about the Internet as early as 1986. As Quarterman notes:

If I may paraphrase, Gore built the Internet in the same way as a mayor builds a bridge: neither by drawing up blueprints nor by welding steel; rather by facilitating its construction. (Quarterman, 1999)

Robert Kahn and Vinton Cerf, who played key roles in the 1970s in coordinating and developing the Internet, wrote a statement where they pointed out that—although 'no one person or even a small group of persons ever "invented" the Internet'— Al Gore was to be given some credit:

As far back as the 1970s Congressman Gore promoted the idea of high speed telecommunications as an engine for both economic growth and the improvement of our educational system. He was the first elected official to grasp the potential of computer communications to have a broader impact than just improving the conduct of science and scholarship. Though easily forgotten, now, at the time this was an unproven and controversial concept. Our work on the Internet started in 1973 and was based on even earlier work that took place in the mid–late 1960s. But the Internet, as we know it today, was not deployed until 1983. When the Internet was still in the early stages of its deployment, Congressman Gore provided intellectual leadership by helping create the vision of the potential benefits of high speed computing and communication. (Kahn and Cerf, 2000)

9.2 RECONSTRUCTING THE INTERNET

It is not surprising that the evolution of the Internet does not fit nicely with existing models of innovation, or that it is not easy to say whether Al Gore should get a patent on it. Exactly because of this, it allows us see where the conventional innovation models fail. One reason for the success of the Internet has been that so many people have felt that they have made important contributions to it. In other words, there are—and there have been—many different Internets.

History is reconstructed from the present, but the present is also constructed using history. We do not only reinterpret authorship and agency retrospectively using hindsight. We also reinterpret the technology itself, based on its current uses and the role it plays in our social practices. This has interesting implications, for example, for policy-making. What, indeed, should we regulate and legislate when we regulate and legislate on 'the Internet'?[2]

A very concrete example of the continuous process of technology reinvention can be found in the recent history of the Internet. Many articles that discuss the Internet at the beginning of the 1990s do not actually discuss it at all. Indeed, the majority of 'Internet' articles from that period—articles that are today categorized in electronic databases using Internet as a keyword—do not mention the Internet. In fact, they discuss cases of paedophilia and crimes where ATM machines play a role. At some point in the 1990s, paedophilia and crime became associated with the Internet, and news articles that discussed paedophilia and computer crime became categorized as Internet articles.

The fact that crime became associated with the Internet is not an accident. Almost all Internet-related articles before 1993 mentioned the famous 'Morris worm' that shut down much of the Internet in December 1988. As a result, the Internet became known as a domain of hackers. The dominance of the Morris worm can be seen in Fig. 9.1. The figure shows both the importance of the Morris worm at the end of the 1980s and the importance of retrospective categorization of news as Internet-related news. The analysed articles include all newspaper articles recorded in the Lexis-Nexis database.

Halbwachs (1980) argued that social order is based on collective memory, which provides the basis for interpersonal meaning. Some ideas fit with the way a social group collectively understands its world. In the case of the Internet, this group is global. The media have retold the story of the Morris worm for over a decade now.

---

[2] Mitchell (1995) and Lessig (1999) have pointed out that technical design choices are becoming increasingly important in regulation. According to Lessig, regulative forces include norms, laws, markets, and technical architectures. The concept of technical architecture, however, implies that we have a functional description of the system at hand. As our interpretation of functionality depends on uses, the concept of functionality makes sense only within a community of practice. This, in turn, means that norms, which are rooted in the internal values of a community, and functionality are both expressions of community values. Lessig's point seems to be that the norms of a user community and the norms of a technology developer community can be independent, and are increasingly so in the modern economy.

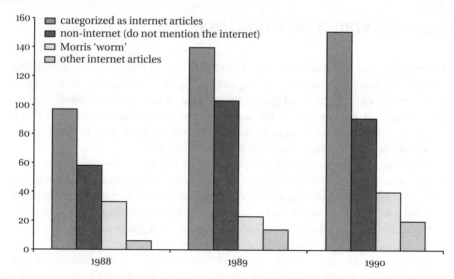

Fig. 9.1. The 'Morris worm' and Internet articles in newspapers, 1988–1990

When anything occurs that has a perceived similarity with this prototypical example, the worm sticks its head above the ground. This can be seen in Fig. 9.2, which shows the number of articles that mention the Morris worm in the Lexis-Nexis database.

Around 1990 the Internet was a network that connected thousands of computers around the world. For many people it was, however, a rather exotic thing: a technological system that had a worm. Most people didn't know about the Internet at all. In fact, the Internet started to attract attention only around the end of 1993. Whereas it was earlier typically described as 'a U.S. computer research network', and around 1990 as a 'rapidly growing global computer network', in 1994 it became increasingly associated with the World Wide Web. Rapidly it became an economic and political topic. This can be seen in Fig. 9.3. The figure shows the number of articles that discuss the Internet. The numbers are corrected so that they do not include articles that have been retrospectively categorized as Internet articles but do not discuss the Internet. Also articles which do not relate to the Internet computer network, are not included in the numbers shown in the figure. For example, many articles that discuss the 'Internet' around 1990 are about the Internet banking system, which does not have anything to do with Internet technology.

The nature of retrospective interpretation of history can easily be seen by studying the content of Internet-related articles in different time periods. For example, the *Financial Times* had 141 articles that mentioned the Internet in 1994. In January 1998 it had 171 articles mentioning the Internet. In the first fifteen days in 2000 it had 484 articles. By iteratively categorizing the topics discussed in these articles a number of themes can be found. For example, whereas about 17 per cent of Internet articles discuss industry-related news in 1994, in 2000 41 per cent of the articles are industry analysis and news. The Internet had also become an investment

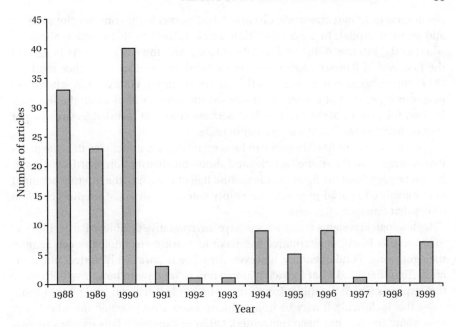

Fig. 9.2. The Morris worm in major newspapers, 1988–1999

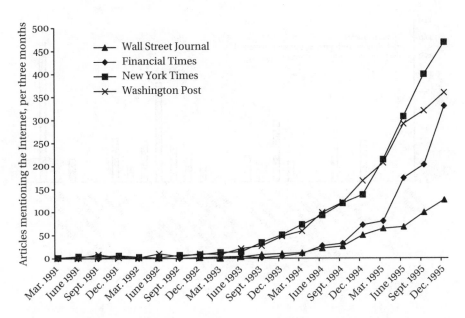

Fig. 9.3. Early Internet awareness

phenomenon. In 1994 one article discussed the Internet in the context of investment and venture capital. In 2000 every fifth article related to these topics. Similarly, whereas the Internet didn't have anything to do with the stock markets in 1994, in the first half of January 2000 almost every third article discussed stock markets. Other topics became non-issues as the Internet evolved. For example, whereas in 1994 over 11 per cent of the articles discussed the number of users on the Internet, in 2000 the growth of the Internet had become common knowledge. Only 1.2 per cent of articles mentioned the number of users.

The emergence of new themes can be seen in Fig. 9.4. It includes the categories that emerge from the studied articles and shows the distribution of articles in 1994. As can be seen from the figure, in the second half of the 1990s, the Internet acquired new meaning in social practice. The empty categories in Fig. 9.4 represent those topics that emerged after 1994.

Technological systems and products have interpretative flexibility. Partly this flexibility results from the alternative social contexts where technologies can acquire their meaning. Fundamentally, however, meaning is always a historical phenomenon. To understand what a specific innovation is, we have to be able to tell a story that explains where the thing came from, who made it happen, and why. In some cases this is difficult. It may be impossible to know who invented the wheel, and how many times it has been reinvented. Often our memory fails us. Lacking historical detail, we can simply invent the missing parts of the story, including the inventor herself. The traditional concept of innovation relies on a teleology where

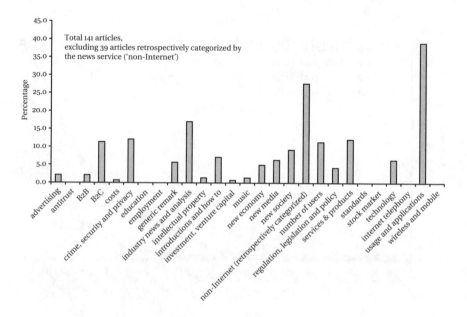

Fig. 9.4. Internet articles in the *Financial Times*, 1994

better functionality is the universal final cause, and from this Archimedean point we start to unfold our stories. In such stories, square wheels are replaced by round ones because of their smoother function. A more historically accurate concept of innovation could be found by noting that there are several actors who have their own perspectives on the various uses and benefits, that there exists a multitude of agents of innovation, and many competing stories that can be told.

10

Learning from Linux

In the Introduction I argued that the Linux operating system is an interesting test case for theories of innovation and technological development. During the last couple of years, the open source development model has been on the front pages of newspapers, and the focus of much attention (e.g. DiBona, Ockman, and Stone, 1999; Wayner, 2000; Raymond, 1998*a*; 1998*b*; Bezroukov, 1999; Kuwabara, 2000; Tuomi, 2001). It has been claimed, for example, that open source projects can produce better technology than traditional corporate R&D (Raymond, 1999). As a result, many corporations have invested heavily in trying to adopt best practices from the open source model.

In this chapter we return to the open source development model and the history of Linux, which were briefly discussed in the Introduction. First, I will describe the Linux system and its developer community in an evolutionary context, highlighting some main characteristics of the socio-technical change that has led to the current Linux system. I will then discuss the organization of this technology creation process, focusing on control and coordination mechanisms. I will describe in some detail the ways the Linux community has managed the trade-offs between innovation and maintainability of the increasingly complex system, and discuss how the learning has been embedded and reflected in the system architecture.

Linux has attracted considerable attention because it has been argued that the open source quality control mechanisms are more effective than traditional methods used in software development. It has often been claimed that Linux is more reliable than proprietary systems because it is developed using the open source principles. I will describe the Linux quality control system, analysing in some detail the bug removal process and the complex socio-technical system that underlies it.

Open source development is a special form of technology development as it intentionally reverses some common intellectual property rights. Instead of copyright it uses 'copyleft', which guarantees the rights of users to modify the results of development, and derive new works from it. The fact that such a licensing model seems to work and promote technology development has important consequences for discussions on intellectual property rights, the patent system, and the theory of appropriation of the results of innovation. The open source licensing policy can be seen as an important social innovation that has a major impact on the way Linux is developed. I will discuss regulations and standards that underlie Linux development, and describe the various forms of licensing that have been used in the open source community.

As was noted before, innovation literature sometimes leaves the process of invention in a black box where undefined psychological forces operate outside the domain of innovation research. The drivers for innovation are commonly understood to be economic. In this context, it is interesting to note the incentives and drivers of technology development as they can be observed in the case of Linux. Reputation and attention are closely related in the Linux community, and they are the key to resource allocation, which, in turn, directs technology development.

The present chapter covers a broad set of issues and introduces some new theoretical concepts. The goal of this chapter is to provide a rich enough description of the history of Linux so that we can see how the concepts presented in the earlier chapters work, and open the area for more detailed theoretical and empirical study.

10.1 THE EVOLUTION OF LINUX

Linux development started in 1991 when Linus Torvalds got a new Intel 386 PC and wanted to learn it. In the beginning, Torvalds didn't expect that anyone would use Linux. It was, however, developed to be compatible with widely used Unix tools, and its source code was made available through the Internet for anyone who was interested. As a result, people who wanted to have a Unix-like operating system on their Intel-based PCs quickly adopted Linux and started to add new functionality to it (Torvalds, 1999).

Linux was inspired by a small Unix-like operating system, Minix, and many of its early adopters were familiar with Minix. Minix had been developed by Professor Andrew Tanenbaum—a well-known authority in operating systems theory—to teach operating systems for students who had only the first generation PCs available. Whereas Minix was intended to introduce the basic theoretical concepts of operating system design, Linux was a more pragmatic project. The goal was to develop an operating system that worked well on Intel 386, and which users were free to modify and play with (DiBona, Ockman, and Stone, 1999: 221–51). The first version of the system was release 0.01, in September 1991. Although it is difficult to find accurate data on the usage of Linux, today there are probably over 15 million Linux users worldwide.[1] Indirectly, almost all people who are connected to the Internet use Linux, as many Web-servers rely on it.

---

[1] http://www.linux.org/info; http://counter.li.org. In 2001 the IDC estimated that about 1.5 million paid copies of Linux were sold in the previous year for client desktops and 2 million copies for server operating environments. The IDC estimate for the annual compound growth rate for Linux server shipments in 1999–2004 was 28.4 per cent (Kusnetzky and Gillen, 2001).

Linux, and its open source development model, started to attract attention around 1994. Until that time the Berkeley BSD Unix had been the most visible open source development activity (McKusick, 1999). It was generally believed that the era of Unix-based operating systems was over, and that Microsoft had secured its position as the dominant player in the operating system market. As an indication of this the Berkeley Unix development group was formally shut down (Raymond, 1999: 22–3). The success of Linux came as a surprise to its developers, but also to people who had been closely observing the evolution of software and open source projects. In his influential article,[2] Eric Raymond describes how Linux made him realize that there exists a new mode in software development:

Linux overturned much of what I thought I knew. I had been preaching the Unix gospel of small tools, rapid prototyping and evolutionary programming for years. But I also believed there was a certain critical complexity above which a more centralized, a priori approach was required. I believed that the most important software (operating systems and really large tools like Emacs) needed to be built like cathedrals, carefully crafted by individual wizards or small bands of mages working in splendid isolation, with no beta to be released before its time.

Linus Torvalds's style of development—release early and often, delegate everything you can, be open to the point of promiscuity—came as a surprise. No quiet, reverent cathedral-building here—rather, the Linux community seemed to resemble a great babbling bazaar of differing agendas and approaches (aptly symbolized by the Linux archive sites, who'd take submissions from *anyone*) out of which a coherent and stable system could seemingly emerge only by a succession of miracles.

The fact that this bazaar style seemed to work, and work well, came as a distinct shock. As I learned my way around, I worked hard not just at individual projects, but also at trying to understand why the Linux world not only didn't fly apart in confusion but seemed to go from strength to strength at a speed barely imaginable to cathedral-builders.   (Raymond, 1999: 29–30)[3]

During 2000 Linux gained credibility as a serious contender for Microsoft. IBM, HP, and Intel, along with other visible partners, created the Open Source Development Lab in Portland, Oregon.[4] Several major software companies started to offer their products for the Linux environment. In 2001 governments around the world launched initiatives to study the open source model and the use of Linux as an alternative to Microsoft operating systems.[5] The Beijing government awarded

---

[2] http://www.tuxedo.org/~esr/writings/cathedral-bazaar/. The article is also included in Raymond (1999).

[3] One should note that historically Raymond's account is not very accurate. The building of medieval cathedrals in many ways resembles the open source development process. Most cathedrals were built over long periods of time as several independent projects and their funding was largely based on gifts (Watson, 1976; Branner, 1961). Visiting a cathedral, it is easy to see that they have historical layers from many different centuries. Notre-Dame was constructed mainly between 1163 and 1250, on a place where previously had been a Roman temple for Jupiter, a basilica dedicated to St Étienne, and a Romanesque church. The Cathedral of Milan was built mainly in 1386–1577 and 1616–1813. Cologne Cathedral took 632 years to finish. During the centuries, wings, towers, windows, and chapels were added, and sometimes churches were built over small older churches. As Branner (1961) notes, very few architects were disturbed by the juxtaposition of old and new.

[4] http://www.osdlab.org/.

[5] One of the most visible governmental studies was conducted by the UK consulting firm QinetiQ (Peeling and Satchell, 2001).

contracts to Chinese open source companies, including the Red Flag Linux. The US National Science Foundation awarded a $53 million grant for building the world's largest virtual Linux-based supercomputer. In January 2002 IBM announced a new series of Linux mainframe computers. Linux and the open source development model had become politically and economically real.

10.1.1 The Basic Architectural Components of Linux

Linux is a fast-growing system. The core Linux—the operating system kernel—consists of software that controls computer hardware and programs that run on it. When new interesting hardware becomes available, the operating system kernel is extended for it. Usually, Linux code for specific hardware components is developed as 'drivers'. Linux is available for several different processor architectures and therefore there also exist several 'ports' of the system.

An operating system can be built based on several different architectures. The architecture of Linux has been strongly influenced by the Unix operating system. Unix architecture is implemented as layers, where each layer provides service to the layer above it (Tanenbaum and Woodhull, 1997). The bottom layer interfaces the software with hardware. A layered operating system can be represented as in Fig. 10.1.

The system kernel is usually considered to be the 'core' of the operating system. It takes care of process, memory, file, security, network, and input and output device management. Utility or system programs are applications that provide key services that are needed for a functional operating system. In a Unix system, graphical user interfaces, user management, command shell, file backup, and, for example, directory listing programs are examples of such system utilities. These programs use the functionality provided by the kernel by calling system functions through the system call interface. The various end-user applications, such as word processors, database management systems, and web browsers, can use both utility programs and direct system calls to interface with the operating system. The operating system kernel, in turn, uses the underlying hardware through hardware-specific drivers that convert operating system calls into function calls that run low-level hardware programs.

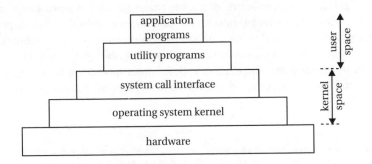

Fig. 10.1. Layers of a Unix operating system

An operating system kernel is not a very useful thing in itself. A fully functional operating system needs system utility programs and applications before it becomes a multipurpose platform that can support and run end-user applications. In practice, Linux relies on a large set of other open source programs to form a fully functional operating system. Most critical of these are the GNU c-libraries and the GNU c-language compiler, which are required for developing the system. A distinction is often made between the Linux operating system kernel itself, and the set of open source applications, including the kernel, that together make a functional environment. The operating system kernel is usually called Linux and the complete system is called GNU/Linux.

The end-users of Linux mainly deal with large software distributions that comprise hundreds of applications in addition to the operating system. For example, the Debian distribution of GNU/Linux has over 1,500 open source programs, including word processors, graphics programs, databases, and web-servers and clients.[6] The evolution of Linux-based systems is therefore only loosely coupled with the evolution of Linux itself. For example, the functionality of GNU/Linux has grown considerably since major application software providers have recently started to port their systems for Linux.

10.1.2 The Growth of the Linux Source Code

A complete Linux distribution is a complex system of interacting software programs. To study the open source development model it is useful to reduce this complexity and focus on the development of the Linux operating system kernel. Already in itself, it provides an interesting example of technology development. Since the first release of Linux, there has been about one new version of the system released every week. During this same time, the total size of the kernel code has grown from 236,669 characters in the distribution files to over 122 million characters. In other words, the code size has grown 516 times.[7] The different versions and their relative sizes are shown in Fig. 10.2. The figure shows sizes for the compressed kernel source code packages. The actual code size is typically about four times larger.

From Fig. 10.2 one can note one of the key characteristics of Linux development. The kernel releases are divided into 'stable' and 'developmental' paths. In practice, the releases are numbered using a hierarchical numbering system where the first number denotes a major version, and the second number gives the version tree in question. In recent years, the even-numbered trees have been stable production releases, and the odd-numbered trees have been 'developmental' releases, where

---

[6] http://www.debian.org.

[7] Code size is measured here in characters, counting comments and documentation as parts of the source. The last available stable release at the end of Jan. 2002, 2.4.17, was used as the end point for the comparison.

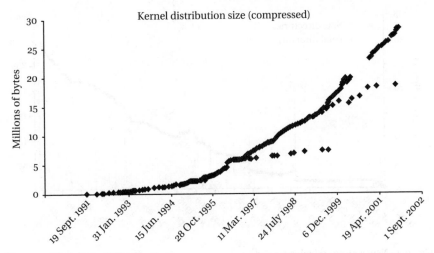

Fig. 10.2. Linux kernel distributions

new features and functionality are introduced and tested. For example, in Fig. 10.2 release paths for versions 2.0.x and 2.1.x create two distinctive paths. The paths fork when version 2.1.0 was introduced in September 1996, and when a new developmental path was started. New releases of the stable path are released in parallel with developmental releases, but usually only with minor bug fixes. Indeed, the last version of the stable path 2.0.x, 2.0.38, was released about three years after the developmental path 2.1 started. The developmental path 2.1, in turn, consisted of 132 versions before it became the next stable version 2.2, at the beginning of 1999.

As Fig. 10.2 shows, Linux development has been active and continuous. In 2000 a major rewrite of the operating system 2.4 was under construction. Originally, Torvalds estimated that the first version of the 2.4.x kernel would be distributed in 1999. The release was, however, delayed for over a year as the kernel architecture was adapted for large files and effective multiprocessor and network use during the development process. During that time, kernel developers worked mainly by sharing patches of code. The resulting gap in Linux releases can be seen in Fig. 10.2. Version 2.4 was finally distributed at the beginning of 2001, making Linux suitable also for everyday operational use in large enterprises.

One of the characteristics of open source software projects is that the system design evolves based on ongoing innovation and learning. One way to illustrate this ongoing innovation is to analyse the increasing complexity of Linux during its history. The structural complexity of the system is reflected in the number of relatively independent subsystems. In practice, the code for each subsystem is organized into its own subdirectory. An estimate of the number of subsystems can therefore be found by counting the subdirectories in the kernel distribution. Fig. 10.3 shows the number of new subdirectories created within two-week time windows, as well as

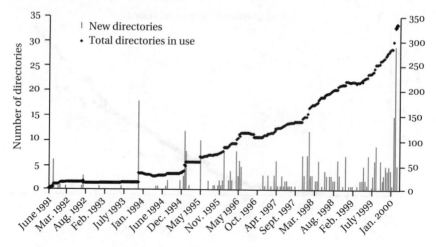

Fig. 10.3. New source code directories

the number of subdirectories in use across time.[8] The left vertical axis shows the number of new directories and the right axis the total number of directories. Major peaks in the number of new directories indicate a major rewrite of the system. This happens when a new major version is released. On average, there were 1.6 new directories created every two weeks. In release 2.3.51, March 2000, there were 333 subdirectories in use. At the end of January 2002, the number of subdirectories had grown to 516 and contained 9,913 files in total.[9]

Using a similar measure, it is also possible to estimate the intensity of 'creative destruction' in Linux development. On a structural level, this can be viewed as the number of system components that become obsolete within a given time window. Using the directory structure as a proxy for this, we can count the number of directories that disappear within a given time window. The result is shown in Fig. 10.4.

Already from this analysis, it is easy to see that the Linux development model has led to continuing system development. Even within the kernel itself, the rate of technology creation seems to increase as the development proceeds. Although the system has gone through a large number of revisions, the rate of growth does not seem to slow down.

[8] The graph was produced by analysing the file creation dates for 11 kernel releases, including 0.01 and 2.3.51, and defining the directory creation date as the date the first file in the directory was created. A moving 14-day time window was used, starting from the first file creation date found. The total number of files was about 20,000. The analysis was done using a set of Perl programs that processed the file lists of the various releases and counted the number of new directories within each time window. This rather labour-intensive process was used as some of the directories in the kernel archives had been recreated during the years, and therefore had lost their original creation dates.

[9] The Jan. 2002 numbers refer to version 2.4.17. The number of subdirectories include all those directories that have files in them. In the following analysis I will focus on changes in the source code before version 2.4. The last version used in the analysis is 2.3.51.

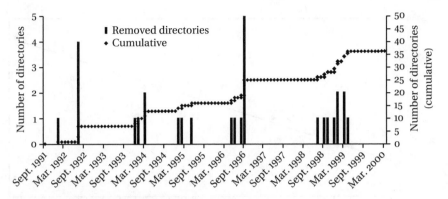

Fig. 10.4. Removed directories as a function of time

10.2 THE LINUX DEVELOPER COMMUNITY

Constant innovation creates major challenges for developing a coherent and maintainable system. When a number of people actively develop the same system, and thousands of end-users can freely report bugs they find and express their ideas for new functionality, there is an ongoing flow of suggestions for improvement. This easily leads to an increasingly complex system that becomes extremely difficult to understand and maintain. In the Linux development community, this phenomenon is known as 'creeping featurism', and it is one of the main concerns of the developers. Yet, it is also important that new innovations are incorporated into new releases without excessive delay. Without the possibility of new contributions being integrated into the system, there would be little point in proposing and producing improvements.

In practice, this inherent tension between the need to incorporate new innovations and keep the system complexity manageable is a critical issue for open source development. A successful resolution of this issue requires effective social coordination and control. The resulting social structures and processes, therefore, reflect the requirements of successful system development. To the extent that Linux is a highly reliable and effective software system, one could then expect that its developer community implements effective social structures for technology development.

Since version 1.0, March 1994, Linux kernel files have included a 'credits' file that lists important contributors to the project. The most recent credits file for Linux contains the names of 407 developers.[10] This is a good estimate of the number of people who have substantially and successfully contributed to the development of the core Linux system. Fig. 10.5 shows the number of people in a sample of Linux credit files.

[10] ftp://ftp.funet.fi/pub/Linux/kernel; the CREDITS file can be found in the root directory of each release. Version 2.4.17.

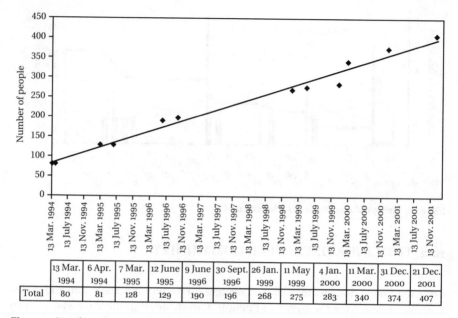

	13 Mar. 1994	6 Apr. 1994	7 Mar. 1995	12 June 1995	9 June 1996	30 Sept. 1996	26 Jan. 1999	11 May 1999	4 Jan. 2000	11 Mar. 2000	31 Dec. 2000	21 Dec. 2001
Total	80	81	128	129	190	196	268	275	283	340	374	407

Fig. 10.5. Number of people in the Linux CREDITS file, 1994–2002

The actual number of co-developers is, however, much higher. There are about 120,000 users who have registered themselves as Linux users,[11] and a large proportion of these have programmed at least minor applications for Linux.[12] These active developers are an important source of bug reports and bug fixes. Often the credit of such contributions is given only in the change logs and in source code comments. The 'bazaar' described by Raymond, therefore, seems to consist of several hundreds of central members, and several thousands of more peripheral but technically sophisticated users.

One important aspect of this 'bazaar' is that it relies heavily on the Internet to get its work done. The Linux development model emerged simultaneously with the explosion of Internet use. In early 1992 it was argued that the development model relied too much on the Internet, therefore excluding people without Internet access (Tanenbaum, quoted in DiBona, Ockman, and Stone, 1999: 245). However, the rapid expansion of Internet use at the time when the Linux kernel was developed provided the developer community with new ways to distribute development work, a new distribution channel, and a global community of sophisticated users.

The regional distribution of early Linux development work is depicted in Fig. 10.6. The figure shows the number of people in different countries mentioned in the first credits file. To adjust for the different sizes of countries, the numbers in Fig. 10.6 are given per million inhabitants. The figure shows that Linux development has

---

[11] http://counter.li.org/.

[12] There were also at least 433 user communities, known as Linux User Groups, in 72 countries in 2001 (http://lugww.counter.li.org/). These typically consist of Linux activists and developers.

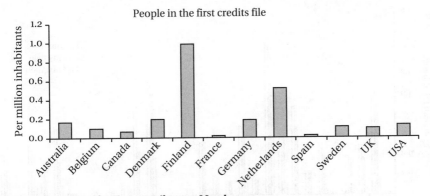

Fig. 10.6. **Location of active contributors, March 1994**

been a geographically broadly distributed activity since the very beginning. The Linux development community, therefore, is a virtual community. It started without face-to-face meetings or the core members ever seeing each other.[13]

The first credits file acknowledged 78 contributors coming from 12 different countries.[14] The credits file for 2.3.51 release, March 2000, had contributors from 31 different identifiable countries.[15] By January 2002 the number of countries had grown to 35. In absolute numbers, the USA was the biggest home base for contributors with 138 people, and 167 contributors of the total 407 came from EU countries.[16] Fig. 10.7 shows the geographical distribution of people in the 2.3.51 credits file.[17]

---

[13] Raymond and others have argued that open source is based on post-scarcity economy and abundance of resources. The evolution of the Linux development community shows, however, that this is not the whole story. For example, according to Torvalds an important reason for distributing Linux through the Internet was the scarcity of collaborators and development resources in Helsinki University (personal communication, Sept. 2000). In this sense, the early phases of Linux were very similar to the early phases of the World Wide Web (Berners-Lee and Fischetti, 1999). As Castells (2001) has noted, resource scarcity also promotes adoption of open source in countries where resources are limited. As these examples show, however, resources are not only economic but also social.

[14] In addition there were two contributors whose location it was not possible to identify using the information in the file.

[15] There were 12 contributors with unidentifiable locations.

[16] The format for entries in the credits file includes the possibility of adding country information but this field is often missing or does not contain country information. Simple text processing of credits files by Perl scripts therefore easily give misleading information. In those cases where country information was missing, a database for contributor countries was built by examining email addresses and other available information and filling some gaps using searches on the Internet. This database was used to fill in missing country information when analysis was done across the different versions. This may create minor errors in the country statistics. Some contributors with no explicit country information are counted as residents in countries where they had lived at some point of time. It is not known how many contributors changed countries while they were working with Linux or how accurately the address information was maintained in the files. For example, of the 407 names in the release 2.4.17 credits file 20 were impossible to map to any specific country. Linus Torvalds is counted as being located in Finland in the early releases and, after moving to California, is included in the count of US contributors.

[17] Luxembourg had one developer in the most recent credits file, but as the country has less than half a million inhabitants, it is omitted from Fig. 10.7.

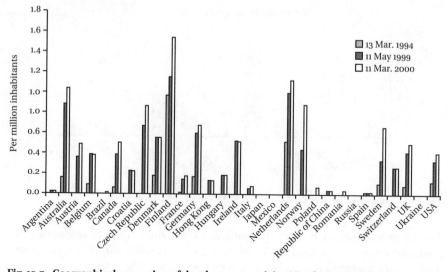

Fig. 10.7. Geographical expansion of development activity, March 1994–March 2000

Linux development is in many ways a self-organizing effort. There is no formal organization, although several non-profit and business organizations have become important in the Linux development effort during recent years. Linux development is in practice organized around projects, communication procedures, communication and collaboration tools, and software modules that are constantly evolving. In many ways, the Linux development community resembles a community of communities, or a fractal organization (Tuomi, 1999), where people are members of a broad 'Linux' community at the same time joining one or more of its subcommunities. These communities are organized around central 'gurus', 'old-timers', and more peripheral novices who have been accepted as legitimate members of the community.

In the case of Linux, the core community members consist of key contributors to the overall kernel development project. In contrast to the basic community of practice model, the Linux development community has more than one centre, as there are several important sub-projects. In recent years, the different main subsystems have been managed by a self-nominated group of maintainers. As a default, Linus Torvalds acts as the maintainer for those subsystems that have no explicitly defined maintainer.

Linux kernel development, itself, is part of a much larger GNU/Linux development activity. For example, the Linux Software Map, which maintains a database of Linux applications and related software, contained 2,120 entries at the end of 2001.[18]

---

[18] The database is available at http://www.execpc.com/lsm/ and at http://www.ibiblio.org/pub/ Linux. The ibiblio-version sometimes contains multiple entries for the different versions of the same software and therefore had almost 5,700 entries at the beginning of 2002. Both databases also have some corrupted entries.

Major Linux-related application projects, such as the Gnome desktop environment project—itself a large umbrella project—are only loosely coupled to the Linux kernel development community.

Although Linux development has very little formal organization, a key to the success of Linux is that development is not random. Development is based on a sophisticated system of social relations, values, expectations, and procedures. This system is in many ways quite different from conventional industrial product creation. It is therefore interesting to see how coordination and control has been managed in Linux development.

10.2.1 Control and Coordination in the Linux Community

Traditionally, organization theorists have argued that increasing complexity in division of labour leads to formal organizational structures (e.g. Mintzberg, 1979). In the case of Linux, this doesn't seem to be the case. Although the Linux community has some structural similarity with the cellular organizational form (Miles, Snow, *et al.*, 1997) and the hypertext organization (Nonaka and Takeuchi, 1995), existing organizational models do not describe very well the structure of the Linux community. Instead, the Linux developer community resembles a dynamic meritocracy, where authority and control are closely associated with the produced technological artefacts. In this sense it also differs from most network-based organizational and innovation models, which typically have focused on firm and industry level networks (e.g. Powell, Koput, and Smith-Doerr, 1996; Van de Ven, 1993; Lynn, Aram, and Reddy, 1997). Indeed, the organizational structure of Linux development could be characterized as a network of communities of practice (Brown and Duguid, 2000), or as a fractal organization (Tuomi, 1999).

A characteristic feature of the Linux development process is its openness. New peripheral additions to the GNU/Linux system are not controlled by anyone. For example, anyone can develop a new application that uses the Linux kernel, and distribute it. As a result, there exists a large set of potential sub-projects competing for community development resources. As will be discussed below in more detail, allocation of these resources is to a large extent based on managing community attention, which in turn relies on accumulation of reputation. Sometimes it is possible to develop a simple program for one's own use and get it added to the Linux distribution; in most cases the development of an interesting subsystem, however, requires that several developers become interested in it. Control, therefore, is indirectly based on capability to mobilize resources. Directly, it lies very much with the users and potential co-developers.

There is no formal organization in the Linux community, but its coordination and control mechanisms can be analysed by observing those explicit and implicit procedures that the community relies on. In the Linux development community, social issues are often described as technical issues. When Linux developers discuss the way the system should be developed and how it should evolve, discussions often focus on code portability, maintainability, possible forking of code-base, programming interfaces, and code size and performance. These technical discussions are critical for the success of the collaborative effort.

Implicitly, each technical choice implies specific procedures and constraints that structure the development work. Often technical decisions are driven by the need to keep the collaborative development going. Computer software is inherently flexible, and there is a very large set of possible ways to implement a specific system. Technical decisions, therefore, are to a large extent articulations of beliefs on the effective ways to organize development. In the course of the evolution of the system architecture, learning on problems and possibilities of collaborative development becomes implemented in the architecture of the technological artefact. Technical discussions on how things 'work', what a good design is, and how development should be done are therefore often reflections of social practices, externalized as specifications for technology. This is most obvious when developers discuss the maintainability of the code, 'cleanness' of the interfaces, and the problem of 'creeping featurism'.

10.2.2 Co-evolution of Social and Technical Structure

A contingency theoretic view would imply that in a successful development project, such as Linux, the structure of software becomes a mirror image of important aspects of the social structure that is needed for a successful system to emerge. Although social institutions, of course, are not directly mapped to the system architecture, the architecture and ways of doing things have to be complementary for a project to be successful. For example, software modularization, coordination mechanisms, incentives, practices of social control, and goals have to be aligned for the overall effort to be successful. These, in turn, have to be embedded within a larger social and cultural context, which limits the arrangements within the Linux community.

The lack of formal organizational structure in Linux development has enabled flexible experimentation with the procedures and values that support effective development. As Linux development occurs in 'Internet time', the speed of evolution is fast. The resulting social innovations, therefore, crystallize some of the learning in organizing collaborative and geographically distributed technology development.

Collaborative software development projects have inherent problems that create specific forms of division of labour, and related design traditions. A sociological description might view emergent social structures in the Linux development community as solutions to underlying social tensions. Blumenberg (1985), for example, argued that social institutions grow around irreducible social contradictions and fundamental conflicts, somewhat as a pearl grows around an irritating grain of sand.

In practice, interdependencies in pieces of software code developed in parallel create a need to coordinate design decisions. Often there are conflicting interests. For example, a minor modification in some part of the program code may require a major rework from people who maintain other parts of the program. A generic way to reduce this problem is source code modularization. A well-designed program has modules whose design limits interactions between modules. If modularization is successful, one programmer can modify the source code of his or her module without requiring changes in other modules. In other words, a programmer can control the evolution of a specific piece of code, without creating problems for other programmers.[19]

In the case of Linux, modularization is based on social agreements, which are supported by commonly accepted development practices, and which are reflected in the overall system design. Many of these social agreements are implicit, and community members have to learn them through socialization. Indeed, only after a novice programmer is able to display the mastery of the key rules, is he or she considered to be a full member of the community. To some extent these rules are also dynamic and they can change.[20]

Sometimes conflicts arise about the implementation and functionality of a specific program module. There can be two competing proposals for architectural choices or two different ways to code a module. If two programmers create different versions of the module, and the module is a key component of the system, this leads to *forking* of the code base. In effect, the system evolves into two different and incompatible variations. This means, in practice, that the synergy in development is lost, and the developers have to choose one of the versions as the basis for their future work. According to Torvalds, such code forking occurred in the first attempt to port Linux to a non-Intel processor architecture. As a result, the kernel design was modified to accommodate new processor architectures in a way that did not risk forks in the code base (Torvalds, 1999: 102).

Design choices for modularization and module interfaces are critical success factors in collaborative software development. It is possible to define modules so that development becomes extremely difficult. For example, if there is no simple mapping between the underlying hardware and the software, the implementation of new hardware functionality may require changes in several modules. Similarly, a small modification in the user interface may require extensive reprogramming if the modularization is bad. The layered abstract architecture of Unix is one attempt to alleviate this problem. In practice, this leads to major challenges in finding the appropriate levels of system abstraction, which are then reflected in the structure of

---

[19] A fundamental problem in the development of complex software is that small modifications in one part of code can have major implications for another part of the code. There is no natural decay in software, and therefore no universal dimension of distance or time. As Wiener (1975) noted a long time ago, digital computers are unique among computational systems because digitalization makes computational machines noiseless, in the information processing sense. Modularity and 'locality', therefore, have to be created and maintained through social processes.

[20] As the Linux development community is a community that develops a functional technical artefact, its social structure has its foundation in the evolving division of labour. In other virtual communities, such as adventure MUDs (multi-user dungeons) and social MUDs (multi-user domains), social rules and control mechanisms evolve when conflicts arise (cf. Reid, 1999; DuVal Smith, 1999). In MUDs, interdependencies between users are limited, and although division of labour is sometimes possible, it is not necessary. The need to functionally link modules makes Linux development a collaborative productive activity where tasks and projects are interdependent.

source code. The situation is made worse by the fact that programmers often want to bypass some levels of the abstract system architecture, usually to improve performance. Often it means that abstract representations of the system only remotely resemble its concrete implementation.[21]

10.2.3 Controlling the Kernel

The constant flow of improvements means that the Linux system is at constant risk of losing its maintainability. In practice, balancing innovation and maintainability has led to tight control of some parts of the system. The control structures, however, are dynamic and continually reproduced in the ongoing communication within the developer community. As Linus Torvalds notes in a recent email:

> If anybody thinks that being the maintainer equals being in 100% control, then I don't think they have understood the TRUE meaning of Open Source. Open source is about letting go of complete control. Accept the fact that other people are wonderful resources to fixing problems, and let them help you.   (Torvalds, 2000)

To study the interplay between control and technology design, it is necessary to describe the internal architecture of the Linux kernel. As was noted before, a GNU/Linux distribution consists of a large set of application programs, Unix utility programs, several versions of the kernel for the different supported processor architectures, and a large number of drivers for different types of hardware. In practice, the abstract layered system architecture that was shown in Fig. 10.1 is therefore relatively close to the actual Linux implementation. On a more detailed level, abstract descriptions, however, start to deviate from the concrete implementation. The kernel, for example, does not have a well-defined boundary between the system call interface and the core kernel. This is partly because there are performance trade-offs, which sometimes make it practical to bypass some internal parts of the kernel. Partly it is simply because the evolution of Linux has led to interactions between the different parts of the system, and, as a consequence, the boundaries have become blurred. Also, in Linux the module called 'kernel', which architecturally most closely resembles the system call interface, implements some process management and memory management functions, as well as some error processing. The main components of the Linux kernel architecture can be represented as in Fig. 10.8.[22]

---

[21] For example, Linux modules that support different networks should in theory be independent modules. In practice, there have been many interdependencies between the modules for different networks. Armstrong (1998) has used automatic architecture extraction tools to analyse the Linux source code, and notes that these interdependencies create a potential maintenance problem for the kernel.

[22] The architecture is extracted from version 2.0.30 (Armstrong, 1998). The dependencies shown in the figure are only some of the main dependencies. As the authors of another Linux architecture extraction study (Bowman, Siddiqi, and Tanuan, 1998) note, automatic extraction is not possible as extraction tools are not able to detect all dependencies and because the Linux kernel modules are highly interconnected. Bowman *et al.* were able to automatically extract about one hundred thousand dependency facts that clustered into fifteen thousand dependencies. Their report provides a useful discussion on the internal architecture of the subsystems.

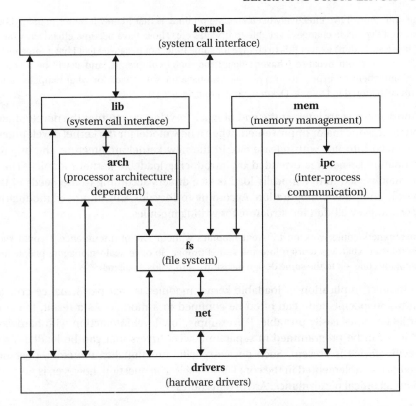

Fig. 10.8. Linux architecture
Modified from Armstrong (1998).

The need to control the operating system kernel was one of the topics in the famous debate between Andrew Tanenbaum and Linus Torvalds in 1992. Tanenbaum argued that it is critical for a successful operating system project that someone maintains tight control of the code, so that its complexity does not explode and so that the core of the system does not fork:

If Linus wants to keep control of the official version, and a group of eager beavers want to go off in a different direction, the same problem arises. I don't think the copyright issue is really the problem. The problem is co-ordinating things. Projects like GNU, MINIX, or LINUX only hold together if one person is in charge. During the 1970s, when structured programming was introduced, Harlan Mills pointed out that the programming team should be organized like a surgical team—one surgeon and his or her assistants, not like a hog butchering team—give everybody an axe and let them chop away. Anyone who says you can have a lot of widely dispersed people hack away on a complicated piece of code and avoid total anarchy has never managed a software project.   (Quoted in DiBona, Ockman, and Stone, 1999: 247)

At that time, Linus emphatically argued that he would not control the system:

This is the second time I've seen this 'accusation' from ast [Andrew Tanenbaum] . . . Just so that nobody takes his guess for the full truth, here's my standing on 'keeping control', in 2 words (three?):
I won't.

The only control I've effectively been keeping on linux is that I know it better than anybody else, and I've made changes available to ftp-sites etc. Those have become effectively official releases, and I don't expect this to change for some time: not because I feel I have some moral right to do it, but because I haven't heard too many complaints, and it will be a couple of months before I expect to find people who have the same 'feel' for what happens in the kernel. (Quoted in DiBona, Ockman, and Stone, 1999: 247)

Almost seven years later, at the end of 1998, Torvalds argued that the development had undergone major improvement when a new model for the kernel development was taken into use with release 2.0. In the new kernel architecture, the original monolithic kernel was extended by introducing loadable kernel modules. These are mainly used to dynamically load device drivers according to the needs of the specific computer configuration. According to Torvalds, this improved modularity by creating a well-defined structure for writing modules:

Programmers could work on different modules without risk of interference. I could keep control over what was written into the kernel proper. So once again managing people and managing code led to the same design decision. (Torvalds, 1999: 108)

An indirect implication of loadable kernel modules is that performance-critical, hardware-specific code can often be confined to a module. As a result, the core kernel becomes easily portable. For example, low-level interaction with hardware devices can be programmed in separate device drivers that can be loaded if the specific device is present. Such device-specific functionality, therefore, does not need to be implemented in the core kernel itself. The question, however, is not only about technical performance. As Torvalds notes:

But Linux's approach to portability has been good for the developer community surrounding Linux as well. The decisions that motivate portability also enable a large group to work simultaneously on part of Linux without the kernel getting beyond my control. The architecture generalizations on which Linux is based give me a frame of reference to check kernel changes against, and provide enough abstraction that I don't have to keep completely separate forks of the code for separate architectures. So even though a large number of people work on Linux, the core kernel remains something I can keep track of. And the kernel modules provide an obvious way for programmers to work independently on parts of the system that really should be independent. (Torvalds, 1999: 109–10)

10.2.4 Balancing Control and Innovation in the Linux Kernel

Comparing Torvalds's early and later statements on controlling the system it is clear that the system development practices had considerably evolved in a few years. In 1992, when there were only few people developing the system, there was no obvious need to restrict derivative works. Although Tanenbaum warned Linux developers about the problems of uncontrolled forking, at that time the developers were more interested in the possibility to easily modify and improve the system. Indeed, the system was perceived as a huge technical opportunity and there were no visible constraints that restricted its future evolution. In other words, it was seen as a platform that could easily adjust to the best technical ideas anyone could come up with. Due to its simple structure, open source distribution, and constant improvements, it was able to effectively grab attention among programmers who wanted to show and use their programming skills, and who enjoyed the possibility of creating code that contributed to the collective effort.

As the system has become more complex and there have been more active developers, the problem has increasingly been in balancing coordination, control, and local innovation. The key factors in this process seem to be modularization and implicit management of attention. Attention is allocated to a large extent based on centrality in the community, and this, in turn, is based on reputation. Reputation within the community is, in turn, to a large extent based on producing working code that has relevance for the community.

Under these circumstances, reputation is a good predictor for future achievements. If someone has successfully coded Linux, he or she most probably knows many things about programming. A distributed system of social control seems to work effectively in this case, and the technical artefact and its developer community evolve in compatible ways. Although the skill-base and tools change, the open and collaborative environment makes it relatively easy to learn new techniques and renew competences. As a result, meritocracy has definite merit in Linux development.

Innovation, however, is also closely related to the modularity and extensibility of the underlying technical system. Effective resource mobilization is possible in the Linux community only because the kernel architecture provides a relatively stable focal point around which dynamic sub-communities can emerge.

Using the directory structure of the Linux kernel, it is possible to illustrate the evolution of the various kernel components. Such an analysis reveals that the growth in the system is quite heterogeneous. Innovation and development are strongly concentrated on some parts of the system, whereas other parts rarely change. The core kernel may be defined as those parts of the system that have stabilized. Fig. 10.9 shows the evolution of some components of this core kernel. A typical pattern is that a period of relatively rapid change is followed by stabilization and lock-in.[23]

The 'core' of the Linux kernel in Fig. 10.9 is defined as those components that have stabilized in the early phases of the development process. This conceptualization means that there is no predetermined categorization of the components, for example, based on theoretical understanding of what are the 'foundational' layers of a typical Unix operating system architecture. Instead, the 'foundational' components are defined as those components that provide a foundation. The fact that these components acquire this role depends very much on the fact that this foundation has to address the needs of several actors. In this sense, foundational components of the structure are its 'institutional' components.

---

[23] The source code growth rate was analysed using a set of Perl programs that processed file size information in a sample of 14 Linux releases, from 0.0.1 to 2.3.51. The growth rate was calculated by dividing by the number of days between releases.

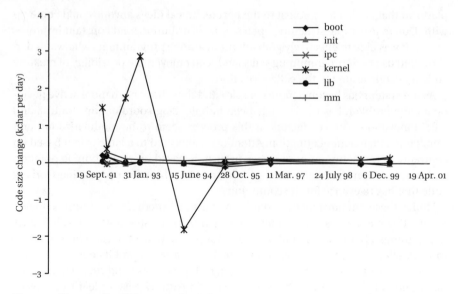

Fig. 10.9. The growth of Linux core kernel components

In most cases, changes in the core kernel would require extensive rewriting of other parts of the system that rely on the functionality of the core. Therefore such changes can happen only in limited ways. In practice, after there exists a substantial amount of code that depends on the core kernel, the core becomes frozen. It can only change its internal implementation or provide new functionality that is compatible with the old.

Technically, however, it is not possible to constrain changes in such a way. In other words, the open source development model needs strong social controls. These social controls are often expressed as design principles and 'good' programming practices. For example, Torvalds notes:

The first very basic rule is to avoid interfaces. If someone wants to add something that involves a new system interface you need to be exceptionally careful. Once you give an interface to users they will start coding to it and once somebody starts coding to it you are stuck with it. (Torvalds, 1999: 105)

The core kernel changes as bugs are corrected or when inefficient code is rewritten. New functionality is introduced only when there are extremely important reasons for it. The conservative policy for the extensions originates from the fear that the core kernel becomes difficult to maintain, or that new bugs are introduced into the core. By keeping the core kernel simple and stable, and by providing support for extensions in the kernel architecture, technical change can be directed to areas where it can be managed. The fundamental trade-off is, of course, that radical innovation in the core kernel becomes difficult, or impossible in practice. The interactions between software modules make the system evolution extremely path-dependent, and tight control of some parts of the system is required to keep other parts of the system open and extensible.

However, as we saw before, Linux has been growing rapidly for almost a decade. As expected, in contrast to components shown in Fig. 10.9, some other parts of the kernel grow very fast. Linux was originally developed for Intel 386 processor architecture, but since version 1.2.0 it has supported several alternative processor architectures. Linux can be extended to a new processor by porting its processor-dependent parts. The code for these different processor-dependent parts is organized into their own directories. Using such a modularization of source code, the code for different architectures becomes independent, and it is possible to add support for new processor architectures without interfering with other parts of the source code. In practice, this has led to a situation where several different teams of programmers have been able to develop the overall system in parallel.

Similarly, the hardware-specific drivers can be developed as independent modules. Indeed, most Linux development in recent years has been related to new hardware components. The open source policy makes it easy for anyone to develop hardware-specific additions to the system, as long as the developer knows the internals of the hardware in question.

Linux architecture is extensible also in areas that one would expect to be parts of the core kernel. For example, Linux supports a large selection of file systems. As long as existing file systems continue to work, it is possible to introduce a new file system without much risk of destroying system reliability. As Torvalds notes:

Without modularity I would have to check every file that changed, which would be a lot, to make sure nothing was changed that would effect anything else. With modularity, when someone sends me patches to do a new filesystem and I don't necessarily trust the patches *per se*, I can still trust the fact that if nobody's using this filesystem, it's not going to impact anything else.   (Torvalds, 1999: 108)

As a result, the evolution of Linux is very much concentrated on those parts of the system that can be developed independently. This can again be seen by analysing the rate of change in the source code size. The main extensible components of the kernel distribution are shown in Fig. 10.10. Comparing the rates of change for core and extensible components of the kernel, one can see that the extensible components grow typically about two orders of magnitude faster than the core components.

In the Linux architecture, institutional innovation—innovation in the core components of the system—seems to be rare. Some other parts of the Linux architecture, however, grow very rapidly. When new hardware is introduced, Linux developers very quickly integrate it with the Linux operating system. Linux, itself, can therefore be viewed as an actor that quickly appropriates new technological elements and turns them into resources for the Linux user community. This is also probably the main difference between conventional software projects and the Linux development project. Linux is clearly an ecology of socio-technical development, not a project that implements a predefined plan. The developmental history of Linux, therefore, also shows us how technological artefacts and social coordination co-evolve. One theoretically interesting position from which to observe some key characteristics of this co-evolution is actor-network theory.

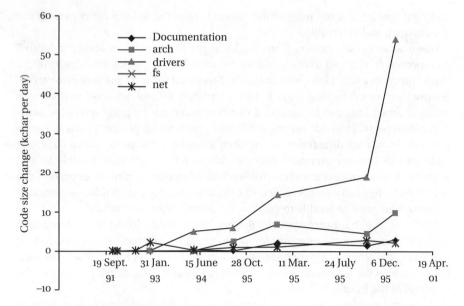

Fig. 10.10. **Growth of architecture-dependent and extensible components**

10.3 SEDIMENTATION, TRANSLATION, AND REDUCTION OF COMPLEXITY

According to actor-network theory, society consists of networks of heterogeneous actors, both human and non-human (Latour and Woolgar, 1986; Bijker and Law, 1992; Callon, Law, and Rip, 1986; Latour, 1999; Law and Hassard, 1999). Society, organizations, agents, and machines are all effects generated through the interactions of actor-networks. A person, for example, cannot be understood as an isolated individual entity; instead, he or she is always linked to a heterogeneous network of resources and agents that define the person as the specific person in question. Without his or her instruments, laboratory, and social relationships, a scientist, for example, loses his or her identity as a scientist. Similarly, a technical artefact can only be understood as an element in a broader network of transformations accomplished by humans and other artefacts. A scientific laboratory may be viewed as a network of test tubes, diaries, scientific publications, budgets, and researchers, each with their own 'competences' and 'resistances'. Scientific knowledge is produced in this network, and becomes an actor itself through new conceptualizations and observations recorded in journals, or, for example, by becoming embedded in scientific instruments and software code. A similar process also underlies evolution of other social institutions. Families, organizations, computing systems, the economy, and technology can all be similarly pictured (Law, 1992: 381).

A key concept in actor-network theory is 'translation'. The total system of actors in the full social network is extremely complicated. Reduction of this complexity is therefore a necessary requirement for practical action. Translation means a process where complicated sub-networks become represented by actants, either human or non-human, and by which the complex underlying structure becomes a 'black box' for practical purposes. For example, sometimes we can talk about 'the British Government', 'post office', or 'email' without having to know what their exact processes are, what technologies they use, or who the people are that make them work. Similarly, an organization can be represented by a single individual, and a complex system of accounting procedures can be represented by a software package.

Translation means that complex sub-networks become 'punctualized', and start acting like a unified entity from the point of view of those actors who interact with the sub-network. At the same time such translated sub-networks become resources. For example, a piece of paper, a pen, a computer keyboard, and an intercontinental missile can be used without considering these processes, knowledge, and other resources that are required to manufacture them. Translation therefore means that complex networks can be taken for granted. But at the same time it means that the point of translation also becomes a locus of power and control. The effects produced by the translated sub-network become resources that can be located and controlled. Through this process of translation the punctualized network can be represented as if it were owned by the actor who manages the translation.

According to actor-network theory, translation generates ordering effects, such as organizations, institutions, devices, and agents. Each of these has its own resistances, and social change therefore is very much about a struggle to reorganize the resources and relations in the actor-network. In this process, resistances are anticipated and various strategies are deployed to overcome them. There is a continuous threat that existing order breaks down, and the fact that order exists indicates that— at least in some pragmatic sense—strategies and translation processes work and form a relatively stable system.[24]

10.3.1 Sedimentation of the Source Code

Linux is a continually evolving technical artefact and is therefore difficult to turn into a resource. One generic strategy for doing this, however, becomes clear when we study the history of Linux. I shall call this strategy 'sedimentation'.

---

[24] Niklas Luhmann (1995) based his theory of social systems on a closely related idea. According to Luhmann, both meaning and social order emerge because complexity needs to be reduced. Meaning, for example, can be defined as order that emerges when one actual interpretation becomes selected from many possible 'latent' interpretations in the cognitive process. The underlying order that makes the world a 'meaningful' world is a network of meaning relations or associations that provide the basis of interpreting the world. Similarly, the specific order that makes fundamentally contingent communicative interaction understandable is what we can define as 'social'. Cognition and communication depend on each other and change as the social system evolves (cf. Tuomi, 1999).

In the early phase of Linux development, its source code was mainly used as a platform for further development of the code itself. When Linux started to be a viable operating system, it became used by people who can be characterized as 'end-users'. For such end-users, Linux was not a complex system of interacting source code modules and programming tools. Instead, Linux became a resource. Furthermore, Linux distributors bundled the operating system kernel with applications and utility programs, and effective distribution required efficient management of software configurations. This created a tension in the Linux development model. For some user-developers Linux was a system where new components were frequently added and which provided interesting opportunities to make novel and high-impact contributions. For such users, Linux remained a complex and evolving network of software modules, function calls, and software procedures. For others, this flexibility was a problem. Continuous change intervened with the translation processes and made it difficult to use Linux as a resource.

As we saw before, as a result of this tension between end-users and developers, the development of Linux has been split into two development paths. The branching of source code into stable and developmental paths can be seen in Fig. 10.2, and in more detail in Fig. 10.11. They show how fundamentally the same product can be

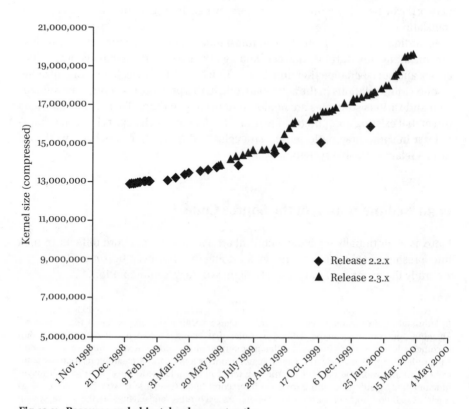

Fig. 10.11. Resource and object development paths

translated into a resource and simultaneously keep evolving as a network. The need to translate the Linux code into a resource for other communities produces a 'sedimented' or 'black-boxed' version of the code.

Sedimentation is a good name for this phenomenon as typically several layers of sediments become formed during the evolution of a system. New code is often developed so that it is 'backward compatible'.

Whereas Linux kernel developers need relatively open access to the Linux source code, and application developers may greatly benefit from such open access, for end-users it is relevant mainly in those cases where the black-box breaks down and reveals its true nature. The proposed 'superiority' of the open source model, there-fore, to a large extent reflects the fact that computer systems often *do* break down. The value of the open source approach is, however, greatly reduced if the end-users do not have enough competence to diagnose the problem. The open source model is effective partly because it also facilitates competence development. More gener-ally, transparency of the underlying system makes it possible for the end-users to mobilize all resources and competences they have available to solve the problem at hand. The specific 'style' in which open source systems break down therefore promotes effective use of problem-solving resources, at the same time facilitating development of competences that can be used in solving similar problems in the future.

To extend the metaphor of sedimentation: open source implies that layers that sediment remain soft. If there is a problem, it is relatively easy to dig one's way through the module interfaces to see where the problem is and how it can be corrected.

10.3.2 Multifaceted Translation and Module Interfaces

According to Schumpeter (1975), the fundamental characteristic of the capitalistic socio-economic system is that resources are dynamically moved from old uses to new ones. Opportunities are materialized in an entrepreneurial process, and the speed of innovation depends on the speed of moving from old activities to new ones. In the history of Linux we can see that such a process also occurs at the level of technological artefacts. Some modules disappear in the course of the evolution, and their developers move to new activities.

As we saw above, a closer analysis of this process reveals, however, that Linux has several qualitatively different 'regions of innovation'. A similar process of sedi-mentation that was seen on the level of the Linux kernel, and which led to the separation of two development paths, can also be seen inside the kernel. The devel-opers of the Linux kernel need to simplify the complexity of the development net-work. Specifically, almost all module developers rely on some key components of the system, which provide core functionality. The translation processes for these key components have to translate the underlying sub-networks simultaneously for many different actors. This is accomplished by sedimenting the representation of, or the interface for, the resource. In other words, the potential problems of main-taining a complex network of changing translation processes is solved by standard-ization of the translation process and by stopping development that could break the black box. The 'hard core' encapsulates the nucleus of the Linux kernel so that Linux developers can continue working on other parts of the system. This is schematically depicted in Fig. 10.12.

The fact that development in these core components slowed down very quickly in the evolution of Linux indicates how difficult it is to provide multifaceted translation interfaces. One might read the rapid stabilization of the Linux core components in Fig. 10.9 as showing that when several different actors approach a sub-network each from their own perspective, no common abstraction is good enough. In other words, there is no generic packaging for changing black boxes. Instead, the code has to be frozen as a concrete technical artefact.

A source code module often acts as a punctualized resource. A standard proced-ure—often implemented as a programming 'interface'—is used to access the ser-vices provided by the resource. As long as the protocol for using the resource and the service associated with it are not changed, the users of the interface don't have to know the internal details of the technological artefact or the organization of its production network.

The use of actor-network concepts, therefore, highlights some generic strategies to create order in socio-technical systems. As we saw before, activity, however, also requires social learning. The community-based view argued that knowing and learning occurs in practice-related communities, and that practices are embedded in material and technological artefacts. In such a context, learning both socializes community members, as Lave and Wenger noted, and creates new forms of activity and new products, as Engeström argued. Actor-network theory, in contrast, argued that human and non-human actors are symmetrical, and that they can often be replaced with each other. The key idea was that the complexity of sub-networks can be reduced by translation and punctualization, which makes one actant able to stand for a whole sub-network.

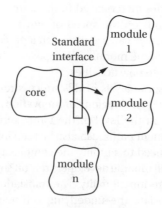

Fig. 10.12. The standard interface as a multifaceted translation mechanism

Putting these two perspectives together shows how both these approaches can be refined and used to describe the evolution of socio-technical systems, such as the Linux kernel. The tools used in social practice are translations of complex sub-networks that produce the tool. The underlying community simultaneously articulates these tools as carriers of knowledge and related practice. As long as technology doesn't break down, its users can use technology as a tool. Such an object is effectively a black box that mediates the user's activity, without requiring the user to consider all the complex relationships that actually are hidden inside the system that makes the tool an object. For example, as long as everything works fine, a computer user doesn't have to know about electrical or digital design, or program architectures, any more than he or she needs to know how these things are developed and produced in practice, or where to find the experts that do know what is inside the box.

In this process of 'black-boxing' sub-networks, translation processes do not only hide complexity of material components. As actor-network theory argued, the network is a heterogeneous one and there are many ways in which material artefacts and humans can switch roles in the network. Black-boxing also hides social networks and discourses. Instead of explicitly negotiating the forms of collaboration, results of earlier negotiations become 'the way we do things here'.

10.4 QUALITY CONTROL, LINUS'S LAW, AND THE ECOLOGY OF BUGS

In the course of Linux's evolution, many new translation mechanisms have been invented. Linux is an exceptionally interesting case of such proliferation of translation mechanisms, as its developers are able to create technical solutions to the problems of translation. In this sense, the Linux community is not only a Linux developer community, but also a tool-developing community. This strength of the Linux development model has been inherited from the Unix culture. Unix was developed as a set of tools that can be easily combined and reused as components of new tools. Indeed, one could argue that this is one of the reasons why Linux development has been rapid. The boundaries between Linux development and Linux tool development activities can be crossed rapidly and without great effort.

The ecology of produced resources is a complex one, and a main challenge in becoming a competent Linux developer is to learn how to use these resources. Some of the resources can be characterized as organizational or community resources, others can be viewed as technological artefacts, or simply as knowledge.

The Linux community has been able to produce high-quality code. Many procedures and tools have been created to manage bugs and to improve the quality of Linux. Product quality is a critical success factor for an operating system project, and it is exactly here that the Linux development model shows its strength.

Almost all software has bugs, and in complex systems bugs can emerge as a result of complex interactions between different program components. Large software systems are therefore difficult to develop. The more developers there are, the more difficult it becomes to control and understand all the possible interactions between software components. This is usually known as 'Brooks' Law'. In *The Mythical Man-Month*, Fred Brooks (1995) noted that adding programmers to a late software project makes it later. He argued that the complexity and communication costs of a project rise with the square of the number of developers, while work done only rises linearly. Raymond, however, observes that if Brooks' Law were the whole picture, Linux would be impossible:

Gerald Weinberg's classic 'The Psychology Of Computer Programming' supplied what, in hindsight, we can see as a vital correction to Brooks. In his discussion of 'egoless programming', Weinberg observed that in shops where developers are not territorial about their code, and encourage other people to look for bugs and potential improvements in it, improvement happens dramatically faster than elsewhere.

Weinberg's choice of terminology has perhaps prevented his analysis from gaining the acceptance it deserved—one has to smile at the thought of describing Internet hackers as 'egoless'. But I think his argument looks more compelling today than ever.   (Raymond, 1999: 61–2)

Quality control is quite a different process in the Linux developer community than it is in traditional product development. Openness means that members of the developer community are able to review the work of others. Whereas traditional software development models often in practice rely on end-users as a source of reclamations and bug-reports, in the Linux model users become problem solvers, providing an enlarged set of problem-solving resources.[25] The user-developers bring many different perspectives, approaches, and experiences into play. From some of these perspectives, a specific problem is easier than from others. When the developer population is large enough, there is probably someone to whom the problem is easy. Raymond formulates this principle as 'Linus's Law':[26]

Given a large enough beta-tester and co-developer base, almost every problem will be characterized quickly and the fix is obvious to someone.

Or, less formally, 'Given enough eyeballs, all bugs are shallow.'   (Raymond, 1999: 41)

---

[25] It should be noted that end-users have played an important role in technology development since the beginning of the computer era. To compete with resource-rich IBM that leased expensive mainframes to its customers, the startup Digital Equipment Corporation sold its machines to its customers encouraging experimentation. DEC, for example, widely distributed documentation about the inner workings of its machines. DEC's first PDP-1 was sold to BBN in 1959, and a number of improvement suggestions came from Edward Fredkin of BBN (Ceruzzi, 1998: 128–9). Due to their open and well-documented structure, the PDP-machines became the core of the hacker culture (Levy, 1984).

[26] Torvalds (2001) himself has formulated another and unrelated Linus's Law. It states that humans have three different types of motives related to physiological, social, and enjoyment needs, that the lower level needs have to be fulfilled before the higher ones become actual, and that human development occurs by rising from the level of lower motives to the higher ones. Although this form of Linus's Law is obviously influenced by the Maslowian theory of needs, Linus, however, seems to think that the progress towards the higher levels occurs through transformation of the motive structure. Torvalds, for example, argues that Linux hackers have so much fun that they can live with a chocolate bar and a Coke.

Raymond argues that this is the main difference underlying the coordinated cathedral-building and the distributed bazaar styles. Whereas development problems are
tricky, insidious, and deep phenomena in the cathedral model, in the bazaar model
one can assume that 'they turn shallow pretty quick when exposed to a thousand
eager co-developers pounding on every single new release'.

And that's it. That's enough. If 'Linus's Law' is false, then any system as complex as the Linux
kernel, being hacked over by as many hands as the Linux kernel, should at some point have
collapsed under the weight of unforeseen bad interactions and undiscovered 'deep' bugs. If
it's true, on the other hand, it is sufficient to explain Linux's relative lack of bugginess and its
continuous uptimes spanning months or even years.   (Raymond, 1999: 42)

10.4.1 Resources in the Bug Removal Process

Raymond's formulation of Linus's Law complements Von Hippel's (1988) argument
that users are important sources of product innovation. Indeed, Linus's Law shows
that there are two essentially different reasons why users are important for innovation. As Von Hippel noted, users can modify and adapt innovations, and thereby
add value to them. As Raymond notes, however, users can also play an important
role in quality control.

This second role is not a trivial one. Indeed, it is a very fundamental phenomenon which has major implications for the theory of innovations also more generally.
Users appropriate innovations in idiosyncratic contexts. These contexts differ both
cognitively and situationally. When the source code is available, software bugs can
be characterized and debugged using these multiple perspectives, each of which
rests on large stocks of unarticulated knowing. In an open source environment,
therefore, software bugs can become opportunities for innovative contribution,
and not just sources of frustration.

In Linux development, the situation consists in part of the complex system of
hardware and software that interacts with the operating system. To limit this complexity, developers, for example, use old and tested compilers to be able to separate
compiler bugs from the kernel bugs. The Linux kernel mailing list FAQ answers
the question of whether different compilers can be used to compile the kernel in
the following way:

Sure, it's your kernel. But if it doesn't work, you get to fix it. Seriously now, there is really no
point in compiling a production kernel with an experimental compiler. Production kernels
should only be compiled with gcc 2.7.2.x, preferably 2.7.2.3. Newer kernels are known to break
the 2.0 series kernels, known symptoms of this breakage are hwclock and the X server
seg.faulting . . . Regarding 2.1 kernels, they usually compile fine with other compiler versions,
but do NOT complain the list if you are not using 2.7.2. Linux developers have enough work
tracking kernel bugs, to also be swamped with compiler related bugs.[27]

---

[27] http://www.tux.org/lkml/.

Often it is impossible to predict the interactions between different system components, and the only way to learn about them is to use the system in different concrete settings. Sometimes the bugs are in the hardware, and there is no way they can be corrected by studying the software. For example, Intel Pentium processors have bugs that need to be corrected by workarounds. In some cases, hardware bugs can be so unpredictable that there is no workaround. The Linux kernel mailing list FAQ lists some of the known processor bugs and, for example, tells that the AMD K6 processor has unpredictable hardware errors:

The AMD K6 'sig11' bug, affects only a few K6 revisions. Was diagnosed by Benoit Poulot-Cazajous. There is no workaround, but you can get your processor exchanged by contacting AMD. 2.2.x kernels will detect buggy K6 processors and report the problem in the kernel boot message. Recently, a new K6 bug has been reported on the linux-kernel list. Benoit is checking into it.

The importance of testing new software code is strongly emphasized, and 'good ideas' rarely get support without a working code that implements the idea. The MAINTAINERS file[28] that lists people responsible for the various kernel modules also gives guidelines for submitting changes to the kernel:

1. Always *test* your changes, however small, on at least 4 or 5 people, preferably many more.
2. Try to release a few ALPHA test versions to the net. Announce them onto the kernel channel and await results. This is especially important for new device drivers, because often that's the only way you will find things like the fact that version 3 firmware needs a magic fix you didn't know about, or some clown changed the chips on a board and not its name. (Don't laugh! Look at the SMC etherpower for that.)
3. Make sure your changes compile correctly in multiple configurations. In particular check that changes work both as a module and built into the kernel.
4. When you are happy with a change make it generally available for testing and await feedback.

The outline of the bug detection and removal process is straightforward. For a software bug to be removed from the system, first someone has to realize that there is a bug. After a bug has been detected, it has to be characterized, preferably by describing repeatable conditions under which the bug can be observed. This phase consists of diagnosing the exact nature of the bug. When the bug has been understood, it can be solved. This phase consists of writing new code that corrects the bug, and testing the new code to verify that the bug has been removed, and that no new bugs have been introduced in the process. When a tested solution is available, it is distributed to other developers. Finally, if the bug is important enough and the new code does not seem to create excessive problems, the bug fix is eventually integrated into a new kernel release. This process is depicted in Fig. 10.13.

---

[28] MAINTAINERS file can be found from the root directory of new releases of the Linux kernel, for example, from http://www.kernel.org/pub/linux/kernel/.

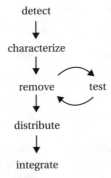

detect

↓

characterize

↓

remove   test

↓

distribute

↓

integrate

Fig. 10.13. The basic bug removal cycle

In actual practice, this rather straightforward process is more complicated. It relies on tools, social practices, and knowledge resources that implement the abstract bug removal procedure. Moreover, the developers apply the various resources in a creative way, improvising according to the needs of the situation. The appropriate way to improvise depends on the audience: if the community of developers understands that a specific way of breaking the standard procedure is justified, rules can be broken. The behavioural standards are usually expectations and suggestions, and there are only few explicit procedures for doing things. Usually such explicit procedures do not result from explicit specification of social processes; instead, they arise from the design of specific tools used in the process. In other words, some aspects of the process are hard-wired into the functionality of the tools.

Some widely used resources and tools for Linux kernel bug management are shown in Table 10.1. The table categorizes the resources as information resources, tools, and communities. Information resources are texts that can be used to learn what the community is doing, what its practices are, and what are the resources available for it. Tools are resources used in the actual bug-removal practices. Community resources are used to keep the community alive and coordinate its activities. As the table shows, one technological artefact can have multiple roles in this ecology. For example, the JitterBug system is a web-based database which shows what bugs are known to the community and whether someone is doing something to correct a bug. JitterBug acts as an information resource by allowing people to find out what bugs are known, and as a community resource by coordinating the work needed to solve the problem.

Some informational resources are meta-level resources that describe procedures used in bug processing. An important meta-level resource is, for example, the linux-kernel mailing list FAQ document that lists frequently asked questions and gives answers and links to further information on them. Some tools interface the object of development, i.e. source code, to the development community. An example of such a tool is the CVS version control system, and the CVS vger–server that maintains the different patches and versions of the kernel in hierarchical trees, and

Table 10.1. *Kernel bug management resources*

Processing phase	Information resource	Tool	Community resource
Detect	Compiled code Documentation	man	LDP
Debug			
Characterize	Source code linux-kernel list FAQ JitterBug oops-tracing.txt Kernel Traffic LDP Project-specific sites linux-kernel archives README files Log files Bug reporting form	Editor gcc make gdb ksymoops IRC Computer configuration	linux-kernel list JitterBug Personal email IRC channels Kernel-newsflash LDP Project-specific lists
Remove	Source code	Editor gcc make	
Test	Patch MAINTAINERS file	diff gcc make Editor ftp	Personal email linux-kernel list
Distribute	Patch MAINTAINERS file	gzip tar email ftp	linux-kernel list JitterBug
Integrate	Patch Release	CVS vger package managers	Maintainers vger

which provides a shared repository of source code to all developers. Many of the tools listed in the table are well-known generic and Unix-tools.[29] Although their existence is often taken for granted, in practice the bug removal procedures critically depend on the tools and their evolution.

[29] *man* is a program for reading manual pages. *gcc* is the GNU c-compiler. *make* is a program that manages the compilation process. *gdb* is the GNU debugger. *diff* is a program that creates difference files from two source code files, and which updates modified files using differences. This is used to distribute *patches* that update files with modifications. *gzip* is used to compress files, and *tar* is used to package several files into one for easier distribution. Linux developers also use generic tools such as *IRC, ftp*, email, and mailing lists. Other tools and resources are systems that are more specific to Linux development. *ksymoops* maintains a list of symbols used in error messages. *Kernel Traffic* is an edited weekly summary of the mailings in the linux-kernel mailing list. *LDP* is Linux Documentation Project, which maintains a set of guidelines and documents for Linux developers. *JitterBug* is system that maintains information of known bugs and patches. *CVS* is a version control system that works with a shared CVS server called *vger*.

Table 10.1 shows some main tools currently used in the bug removal process. One should note, however, that many of these tools have emerged during the evolution of the kernel. Some of the tools and resources explicitly address problems that the success of the kernel development has created. For example, the Kernel Traffic list[30] provides an edited summary of the large volume of mailings in the linux-kernel mailing list. The linux-kernel mailing list FAQ, in turn, documents common questions that novice developers have, as a way to keep such relatively low-priority questions from crowding the linux-kernel list. Similarly, the mailings in the linux-kernel list are archived, so that they can be searched when someone needs to know whether something is known about a potential bug. In that way the mailing list archives provide a simple but effective form of community memory.

Already a superficial analysis of the tools and resources used in the bug removal process reveals that a complex socio-technical system underlies this apparently simple process. Quality control in the Linux kernel, therefore, is not only about finding bugs and correcting them. It is also very much about the complex and continuously evolving system that makes the detection, characterization, and removing of bugs possible in the first place.

Tools and resources therefore mediate the relations between developers, the developer community, and the technical object that is developed. The overall bug removal system can then be represented in a simplified conceptual way as in Fig. 10.14. Raymond (1999) emphasized the cognitive capabilities of co-developers. Sociocultural analysis (e.g. Wertsch, 1998; Leont'ev, 1978; Engeström, 1987; Cole, 1996) would highlight the fact that cognition also very much depends on the tools and resources that are available for the developers. Implicitly, the guidelines for kernel bug removal note this when they insist that new patches need to be tested in different hardware configurations. Linus's Law could then be augmented by noting that a combination of eyeballs and other resources makes even the most insidious bugs shallow.

Quality control in innovative and continuously evolving projects is essentially about learning. Whereas the traditional models of learning in product development focused on decreasing errors in a given product design, in the case of Linux learning is also creative. Theoretical models of innovative learning generally claim that learning starts when a problem arises, and innovative solutions are generated in the process of defining ways to overcome the practical problem at hand (e.g. Dewey, 1991; Schön, 1963; Engeström, 1987).

The Linux development model is compatible with such theories of innovative learning. In this sense, it is also different from the conventionally used product development practices (cf. Griffin, 1997; Mahajan and Wind, 1992), which rarely consider the microstructure of learning. The Linux model, however, does highlight some characteristics of successful product development that have been discussed within the disciplined problem-solving literature on product innovation (Brown and Eisenhardt, 1995). Within this literature, the importance of exploratory learning,

[30] http://kt.zork.net.

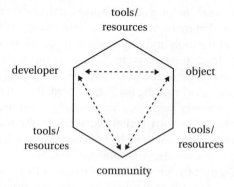

Fig. 10.14. Mediated interactions in the bug removal process

non-financial goals, continuous problem-solving, and diversity of problem-solving resources have often been noted.

10.5 RULES, REGULATIONS, AND INTELLECTUAL PROPERTY

As was noted above, procedures that underlie Linux development are often learned when novice developers become socialized into the community. Many of the procedures and practices are also embedded into the functionality of the tools that support the development. There also exist, however, important explicit standards and agreements that are key components in the development system. On a technical level, one such standard is the ISO Posix interface standard, which defines the way application programs can use the kernel functions.

A distinctive characteristic of open source projects, when compared with traditional corporate software development projects, is the way intellectual property rights are handled. One key innovation in open source has been the GNU General Public License (Stallman, 1999). This has made it possible to legally improve and adopt software developed by others, at the same time facilitating continuous improvement.

The open source licensing rules are important for learning and for creating derivative works. Open source licenses, however, also have important implications for the way resources can be used. For example, the c-language compiler used in Linux development is produced by the GNU gcc community, which is not part of the Linux development community. The gcc compiler is a critical resource for the Linux community (Torvalds, 1999). If the compiler were to become unavailable, Linux development would become difficult or impossible.

The GNU General Public License obviously plays an important role here. It guarantees that the gcc compiler can be appropriated by the Linux community as a core resource in the development. By relying on the institution of copyright, open source licences provide an institutional basis for trust (Kramer and Tyler, 1996). Without open source licence, it would be very risky to build a system that so critically depends on a resource that is produced outside the community. Commercial software development projects have to manage these types of risks by constructing complex legal arrangements and by relying on socially institutionalized sanctions. Often software developers simply have to go for the 'least risky' choice by selecting the biggest vendor. Using a standard interface to the intellectual property institution, Linux developers avoid this additional and often costly translation to legal practice, and this is probably one important reason why change in the ecology of Linux development is rapid.

Copyright law itself is here appropriated as an unintended resource, and used to 'encapsulate' software so that it can become a resource for many different actors, without the need for the actors to negotiate and coordinate their actions. Deployment of this standard translation mechanism avoids the increase in complexity when new actors start to use these translations and the underlying subnetworks as resources in their own activities. This strategy for reducing complexity by defining such a 'property interface procedure' is exactly the same as that which generated the idea of utilizing interface message processors in the ARPANET architecture. Whereas the IMP hardware made it possible to develop simple host-to-host protocols, the GPL licence terms make it possible to create a complex network of software without extensive cross-licensing.

In the history of software, open source models have, however, been used before copyrights became an issue. For example, as was noted before, about half of the operating system programs for CTSS, an early time-sharing system at MIT, were developed by the users of the system (Fano, 1967). One of the motivations for launching the ARPANET project in the 1960s was the belief that by connecting different computing sites, communities of computer programmers could more efficiently share their programs and knowledge. Indeed, two of the most influential visionaries of ARPANET, J. C. R. Licklider and Robert Taylor, argued in great detail in 1968 that such on-line communities would radically transform computer programming, but also society, work, and human thinking. Although they saw security and privacy as important challenges in on-line communities, their underlying assumption was that—within a given access control policy—software could be freely used and shared (Licklider and Taylor, 1968).

Copyright—and intellectual property more generally—is a social institution that reflects history. During this history, some interests have become central and others have disappeared into oblivion (Machlup and Penrose, 1950). As a social institution, copyright helps social actors to coordinate and control their interactions. In open source development projects, copyright helps solve a well-known problem. Access to source code facilitates learning, improvement, and integration with other systems. One important function of copyright agreements is that they keep this development path open, in a world where innovation is increasingly embedded in software (Lessig, 1999), and where commercial appropriation of research and development investments has become increasingly difficult (Davis, Samuelson, Kapor, and Reichman, 1996).

10.5.1 Forms of Open Source Copyright

There exist several variations of commonly used open source licence policies, some of which are more restrictive than others. In a clear contrast to the typical use of copyright licences, which restrict the ways the copyrighted work can be used, the main goal of the free software licences is to guarantee the ongoing reuse and development of software.

In commercial software, the license terms are designed to protect the copyright. They're a way of granting a few rights to users while reserving as much legal territory as possible for the owner (the copyright holder). The copyright holder is very important, and the license logic so restrictive that the exact technicalities of the license terms are usually unimportant . . . In free software, the situation is usually the exact opposite; the copyright exists to protect the license. The only rights the copyright holder always keeps are to enforce the license and to change the license terms of future versions. Otherwise, only a few rights are reserved and most choices pass to the user. In particular, the copyright holder cannot change the terms on a copy you already have. Therefore, in free software the copyright holder is almost irrelevant—but the license terms are very important.[31]

Free software licences guarantee various rights to use, modify, distribute, and distribute modified code. According to the Debian Free Software Guidelines, and the Open Source Definition (Perens, 1999)[32] that has been derived from it, there are several requirements that a software component must meet.[33] First, the licence must guarantee that the code may be freely distributed without royalties.[34] Second, the source code must be easily available, and the licence must not restrict the distribution of the source code. Third, the licence must allow distribution of modifications and derived works under the same terms as the original code. These are the main characteristics of open source software. In addition, to comply with the Debian Guidelines and Open Source Definition, the licence may restrict distribution of modified source code only if it allows distribution of 'patch files' that can modify the original code at the compile time. This is to simultaneously guarantee that the original programmer can maintain the integrity of his or her code, and that subsequent modifications are still possible by adding new 'patches'. In addition, the licence must not discriminate against any persons or groups, or against any uses, including commercial use. The licence must also apply to all to whom the program is distributed, without the need to write separate licence agreements. Further, the licence must not require that the program be used as a part of a specified software

---

[31] 'Free software licensing alternatives'. http://metalab.unc.edu/pub/Linux/LICENSES/theory.html.

[32] The Open Source Definition is available at http://www.opensource.org/osd.html.

[33] 'A social contract'. http://www.debian.org/social_contract.html.

[34] There was a heated discussion going on at beginning of 2002 as some members of the World Wide Web Consortium (W3C) wanted to introduce royalty payments for patents related to Web standards. The proposed RAND (Reasonable and Non-Discriminatory) licensing mode attempted to align the interests of intellectual property owners (corporations paying W3C membership fees) and the development of the Web. In practice, the RAND licensing scheme might make it impossible to develop open source software based on such standards (cf. http://www.w3.org/TR/2001/WD-patent-policy-20010816/).

distribution. To avoid contamination of licences, for example by requiring that the program be distributed only together with other programs that have similar licences, the licence must not place restrictions on other programs.

The least restrictive form of licence is public domain, which puts no restrictions on the use or distribution of the original code. It can be freely copied, used, and modified for any purpose. If a public domain program is available as source code, it adheres to the Open Source Definition.[35] A rough estimate of the use of public domain licences is that in mid-1997 about 3 per cent of about 2,600 software packages and documents on the Sunsite server were defined as public domain sources.[36] Public domain licences are therefore not very common within the open source community.

The least restrictive commonly used licence is the MIT or X consortium licence, which requires only that the original copyright and licence terms are included in the distribution. Shareware programs often use this type of licence, although they may also request a donation from users who find the program useful.[37] A slightly more restrictive licence is the BSD-licence, which requires that all documentation and advertisements acknowledge the original copyright holder. Freely Redistributable Software, in turn, has an FRS licence, which requires that software can be freely copied, used, and locally modified. It must also grant the right to distribute modified binaries, although it can put some restrictions on the ways the modified source code can be distributed. To be 'open source', FRS restrictions have to adhere to the Open Source Definition, however.

The most widely used free-software licence is the GNU General Public License, or GPL. GPL was originally defined by the Free Software Foundation, with the explicit aim to promote non-proprietary software. GPL proponents argue that proprietary software limits innovation, and that fair use of software should be allowed in the same way as fair use of scientific results (Stallman, 1999). This is the licence under which the core Linux system is distributed. It allows free copy, use, and modification. Modified source code can be redistributed if the modified source code shows a 'prominent notice' of the modification. The GPL licence also requires that if a program contains components that are licensed under GPL, all the components must have a GPL. This last requirement of GPL has no simple interpretation in practice (Perens, 1999). To enable commercial programs to be developed for the Linux platform, the licence in Linux explicitly declares that the use of the system is not considered to generate a derivative work. This means that commercial and proprietary programs can use Linux even when they don't want to use GPL. The original idea in GPL was that it shouldn't be possible to make open source software proprietary by adding to it some proprietary components (Stallman, 1999).[38]

---

[35] Many programs have also been put into the public domain without distributing their source code. It is, however, difficult to know what such PD programs actually do if there is no source code available.

[36] http://metalab.unc.edu/pub/Linux/LICENSES/theory.html.

[37] The term shareware is also often used for programs that are distributed without their source code.

[38] Richard Stallman (1999), the founder of the Free Software Foundation, has noted that the recent open source movement has to a large extent neglected the ethical implications of software licensing,

Linux Software Map (LSM) is a database of software available for the Linux platform.[39] Each entry in the LSM includes a field for describing the copyright policy for the software package in question. By analysing the contents of the LSM database, it becomes clear that the GNU licensing policy has been the most common policy among Linux developers. About 61 per cent of the entries had a GNU-type 'copyleft' licence policy. The distribution of licence types in the LSM database at the beginning of 2002 is shown in Table 10.2.[40]

During the history of Linux, developers have also learned that licences are important. Whereas it was quite common until the mid-1990s for developers to describe the copying policy for their software in ambiguous terms, towards the end of the decade the use of copyrights became more exact. This can be seen from Fig. 10.15. The GNU-type 'copyleft' licences have been common for the whole history of the Linux kernel development and more widely used than all the other well-defined open source licences.[41]

A key factor in open source development is that formal contracts are intended to promote development, not to restrict it.[42] Property rights are used here to enable

and focused on a short-sighted way to the productivity aspects of the open source model. Stallman and Raymond are often presented as two ideologically opposite poles of the open source movement. Whereas Raymond emphasizes the effectiveness of the open source development and argues that libertarian values are compatible with the open source model, Stallman emphasizes common good and social responsibility. The Raymondian open source activists sometimes argue that Stallman and the Free Software Foundation is based on 'ideology' and that they are antagonistic to free markets. Such rhetoric, of course, assumes that free market and libertarian values would be free of ideology. Stallman, himself, has often emphasized that when he talks about 'free software' he talks about it in the sense of 'free speech', not in the sense of 'free beer'.

[39] The Linux Software Map was created by Jeff Kopmanis, was taken over by Lars Wirzenius, and is now maintained by Aaron Schrab at http://www.execpc.com/lsm/. There is another popular version of the LSM at ibiblio (http://www.ibiblio.org). The ibiblio database contained 5,683 entries at the beginning of 2002 whereas Schrab's version contained 2,120 entries. The main reason for the difference is that the ibiblio database has multiple entries for the different versions of the same software. The ibiblio database also has many corrupted entries.

[40] The entries in the LSM database do not describe copying policies in any well-defined way. For example, some entries declare their copying policy as 'public domain excluding Microsoft', 'do not sell it', 'free for people friendly to the USA.', 'freeware but send a postcard', etc. Many entries also refer to the copyright rules described in the documentation of the various software modules comprising the package. Many packages also include modules with different copyright policies. For the purposes of the present analysis, the 2,120 entries of the LSM were formatted by a Perl program, imported to a Lotus Notes database, and categorized using the Notes full text search tools. Entries that had multiple well-defined copyright policies were categorized as 'copyleft' if they included GNU-type licences. The free format entries were then categorized using information in the entries. For example, 'do not sell it' was categorized as 'open with restrictions, non-commercial use', and 'free for amateur radio use' was categorized as 'open with restrictions, other'. The copyright policy for some entries was checked from their original distri-bution sites. As the terms used to describe copying policies have changed during the years, it was difficult to categorize entries for shareware and freeware, and to know whether source code was included with the package or not. The figures in the table, therefore, give a good overview of the licence policies used for the LSM entries, but have some room for interpretation in particular for those entries that don't refer to well-defined licence policies.

[41] The figure was generated by a Java program that analysed the entries in the Lotus Notes database described in the previous note. Starting from the first entry date in the database, and using a 183 day time window, the entries in each policy category were counted. The number of entries in each time window is shown in the figure.

[42] Open source projects therefore also show that there are intellectual properties for which appropriation of returns on investment is not a major issue. In the historic controversies on patent rights

Table 10.2. *Licence types in the Linux Software Map database, 2002*

Licence type	Entries
Open source copyleft (GNU, 'copyleft')	1,285
'Freeware' without explicit licence	334
Open source copyright (BSD, MIT, FRS . . .)	145
Open with restrictions	142
Public Domain	77
Shareware	58
Proprietary (commercial, demo)	35
Free distribution, no source	3
Documentation	5
Not defined	36
Total	2,120

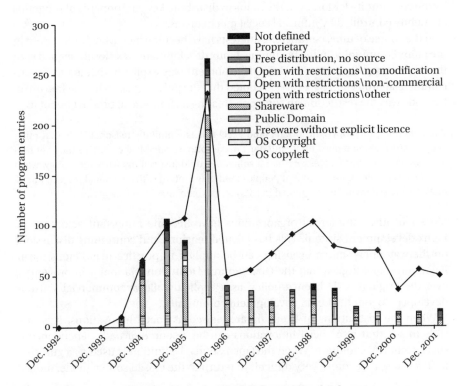

Fig. 10.15. **Licence type distribution in the Linux Software Map database, 1992–2002**

symbiotic development, instead of competition. The fact that innovative use of property rights can provide a platform for effective collaboration and division of labour also without economic competition is an important insight of the free software movement. It also shows that pure market and competition-based analyses of technology development miss a fundamental aspect of socio-technical change.

10.5.2 Social Contracts

The copyright licence, although important, is only part of the story, however. Several explicit and implicit expectations define appropriate behaviour within the open source community. For example, the Debian GNU/Linux community has defined a 'social contract' that declares its commitment to keeping the programs free software, transparency in handling software bugs, and support for users who develop commercial and restricted software based on the free software developed by the community.[43] Moreover, the rights to distribute key components of programs are tightly controlled by informal social mechanisms.

In the course of time, commercial interests have become increasingly important in the Linux community. In the beginning, Linux development was closely aligned with the free software movement. Linux development was explicitly defined as a non-commercial project. In 1992 Torvalds noted that the only exception for the free use of the code was the restriction on someone creating a commercial product out of it:

The only thing the copyright forbids (and I feel this is eminently reasonable) is that other people start making money off it, and don't make source available etc . . . This may not be a question of logic, but I'd feel very bad if someone could just sell my work for money, when I made it available expressly so that people could play around with a personal project. I think most people see my point.   (Quoted in DiBona, Ockman, and Stone, 1999: 248)

More recently, commercial organizations have become important actors in the Linux development system. This has created tensions and continuing discussions on the way open source licensing can be applied in practice (e.g. Perens, 1999). Raymond (1999) argues that the Open Source Definition is a major improvement over the original GNU licence policy, as it explicitly allows commercial software developers to join the Linux development community.

When the institutions of licensing are viewed as social innovations, it is possible to see that also social innovations can be a source of path dependence in socio-technical evolution. When the Open Source Definition is used as a guideline for licensing, it becomes very difficult to return to the closed source mode. Indeed,

(Machlup and Penrose, 1950), the proponents of a free market argued that patent rights may slow down development as they distort markets and do not necessarily allocate returns to those who contributed to the invention. Both free market advocates and proponents of patent monopolies, however, missed the possibility that technological development can result from giving away monopolies.

[43] http://www.debian.org/social_contract.html.

this was exactly the intention of the Free Software Foundation when it designed the GNU licence, with the aim of guaranteeing that the results of technical work can accumulate. The Open Source Definition, with its less contagious licensing policy, however, makes it possible to incrementally develop closed extensions to the Linux system. In practice, this may be difficult as most developers rely on the collective resources of the community, and unfair free-riding easily leads to social exclusion. The transparency of the open source development model also means that it is difficult to hide such attempts at free-riding.

An important role of copyrights and explicit declarations of expectations is that they create norms that act as interfaces between developer communities. They reduce the risks of relying on resources produced by others, and create an environment where trust can be accumulated and transformed from calculus-based trust to knowledge- and identification-based trust (Lewicki and Bunker, 1996). If all resources were to be generated within a given community, social learning and the internal power structure within the community would in most cases make it unnecessary to explicitly declare copyrights, for example. Copyright, however, institutionalizes relationships between communities. By using the GPL, the GNU gcc developer community guarantees that the Linux developer community will have access to this critical resource. But whereas copyrights and intellectual property are often viewed as the basis for economic transactions of knowledge related goods, or 'knowledge economy', open source communities use them to manage an economy where trust, reputation, competence, and attention operate in an ecology of productive communities. In this economy, money is often perceived as a technology which distorts effective allocation of resources.

10.6 DEVELOPER INCENTIVES AND RESOURCE ALLOCATION

Linus's Law and the alignment of social and technical structures may explain why community-based software development can lead to high-quality results. Theoretically, Linus's Law means that each contributor can contribute where his or her impact is greatest. In this sense, the Linux development community implements a market where cognitive resources are effectively allocated.

The existence of such allocation mechanisms do not, however, explain why the product emerges. In the case of a software project, some development may occur simply because debugging, coding, and solving technical problems can be rewarding as such. This, indeed, is an important driver for development. Developers often describe the joy of hacking as their primary motive (e.g. Raymond, 1999; Torvalds, 2001).[44]

---

[44] Torvalds (2001) and Himanen (2001) have tried to explain the motivations of Linux developers using the popular Maslowian hierarchy of needs. Maslow's basic idea was that human needs form a hierarchy in which lower-level needs have to be satisfied before the higher-level needs become

To be able to make relevant contributions to the project, one has to skilfully use tools and concepts, and do something that no one has ever done before. If one succeeds in creating a new piece of software that is taken into use in the community, there is clear evidence of success and a socially validated proof of mastery. Indeed, software projects provide unlimited opportunities for testing one's skills and creating new grounds for mastery. In this sense, one may regard 'the joy of hacking' as something highly non-trivial. Instead, it can be seen as a prototypical driver for technological progress. As was noted before, Csikszentmihalyi (1990) argued that people are most happy when they are performing on the edge of their competences. To an important extent, software developers live in a world of their own creation (Weizenbaum, 1984). In such a world, each new advance and border-crossing moves the boundaries further, expanding the domain where new achievements can be realized. In this context, Linux, therefore, is not just an operating system kernel, but an interesting metaphor of modern technological culture.

Such socio-cognitive explanations are important parts of the whole picture when we try to understand the drivers of technological change. As such, however, they cannot explain the fact that a technological system evolves as a coherent system.

Raymond proposed that the dynamic of Linux development can be understood by noting that the ownership rights that underlie the development are essentially similar to those that underlie the Anglo-American land tenure (Raymond, 1999).[45] In this Lockean theory of property rights, ownership can be gained in three different ways. First, in frontier areas that have never had an owner, one can acquire ownership by homesteading: by mixing one's labour with the unowned land, fencing it, and defending one's title. Second, in an area where ownership already exists, one can acquire ownership through the transfer of the title. In theory, at least, such a chain of title goes back to the original homesteading. Third, property that has been abandoned can be claimed by adverse possession. This happens in a similar way as the original homesteading: one moves in, improves the property, and defends the title as if homesteading (Raymond, 1999: 93). Similarly, in the space of potential technological developments of the Linux system, developers can gain ownership rights for specific sub-projects.

These informal ownership rights are important because they make exchange possible. According to Raymond, the exchanges that underlie the success of Linux, however, are not conventional economic transactions. Instead, following Rheingold (1993) and others (Kollock, 1999), he suggests that the system of social exchanges can be understood as a gift culture (Raymond, 1999: 97). The developers give the results of their work as gifts to the community, and the mutual exchange of gifts leads to a technically highly advanced system with a very high quality. By giving

actual. In the Maslowian context, software hacking, therefore, becomes a form of human self-actualization and gains an aura of human emancipation. One should, however, note that there is no empirical support for Maslow's hierarchy or the different categories of needs proposed by Maslow (Wahba and Bridgewell, 1976; Soper, Milford, and Rosenthal, 1995). Maslow's conceptualization of 'needs' is also theoretically highly questionable (cf. Harré, Clarke, and De Carlo, 1985).

[45] Raymond discusses the property rights in his article 'Homesteading the noosphere'. This is included in Raymond (1999), and also available at http://www.tuxedo.org/~esr/writings/.

gifts, the developers are also able to build reputation. Good reputation among one's peers is a source of reward in itself. Reputation, in turn, makes it easier to mobilize community resources. In some cases, good reputation within the community may spill over to other areas of society, and earn a higher status there.

In this sense, the Linux development community is similar to academic disciplines (Merton, 1973). As Raymond notes, one peculiarity of such communities is that only the members of the community can appreciate the quality of gifts. Indeed, in the Linux community the value of a gift is what others can make out of it.

There is, however, more than one way to run a gift culture. According to Raymond, two sides of gift culture are represented within the software development community by crackers, who try to gain reputation by breaking computer security, and by benevolent hackers, who gain reputation by sharing useful software in source code (Raymond, 1999: 100). The cracker culture is a tightly closed one, and protects its secrets, whereas the hacker culture is based on transparency and openness. This has obvious implications for the way competence, knowledge, and technological artefacts develop. Openness means that results and techniques can accumulate, as it is relatively easy to learn from others' work and add to it. There is a very strong expectation within the community that developers develop their systems in ways that make it possible and easy for others to improve on them (e.g. DiBona, Ockman, and Stone, 1999: 221–51). This expectation is reflected, for example, in the Open Source Definition, which forbids deliberately obfuscating source code, and which requires that source code be distributed in a format that a typical programmer would use to modify the program.

Raymond argues that Linux development works well because reputation is mainly associated with software modules. Although, according to Raymond, developers are driven by ego-satisfaction, there are strong taboos on claiming personal credit. Reputation is made objective by associating it with the produced technical artefacts. Although hackers relatively freely flame each other over ideological and personal differences, it is rare that they would publicly attack someone else's technical competences. Instead of criticizing each other, they criticize the software.

Bug-hunting and criticism is always project-labeled, not person-labeled. Furthermore, past bugs are not automatically held against a developer; the fact that a bug has been fixed is generally considered more important that the fact that one used to be there.   (Raymond, 1999: 110)

Raymond also notes that the hacker culture consciously distrusts and despises egotism:

... self-promotion tends to be mercilessly criticized, even when the community might appear to have something to gain from it. So much so, in fact, that the culture's 'big men' and tribal elders are required to talk softly and humorously deprecate themselves at every turn in order to maintain their status.   (Raymond, 1999: 107)

In Raymond's terms, reputation is very much 'project-based'. His interpretation is that most hackers, as members of the cultural matrix, learn that desiring ego satisfaction is bad. However, he also notes that the rejection of self-interest in the hacker community is so intense that it probably plays some other valuable function.

Raymond proposes two explanations for the taboos on posturing and personal attacks on technical competences. First, when results are judged by their merit, the community competence base increases rapidly. The taboo against ego-driven posturing therefore increases productivity. More importantly, however, when personal status is discounted, the community information on system quality closely reflects the quality of the system, and does not become polluted by personal reputations of the developers.

Implicitly Raymond's account on the reputation mechanisms assumes that there are two systems of reputation operating in parallel. One drives the developers as seekers of ego-satisfaction, whereas the other describes the quality of produced software artefacts. Meritocracy, in the Linux developer community, assumes that there exists a straightforward mapping between the quality of software and the reputation of its developers. As Merton and others have shown, the allocation of individual reputation, however, is a complex social process and often only loosely coupled with the quality of results associated with the individual in question.

The topic of reputation has been very popular in recent years, especially among economists working in the game theoretic tradition. Economists generally assume that reputations are based on information on individual behaviour (Wilson, 1985) and that their role is to make social relations less risky (Granovetter, 1985; Kollock, 1994). Although Raymond's account on reputation associates it also with material artefacts, fundamentally he locates reputation among characteristics of a specific individual. This may appear to be a natural and obvious choice. The usefulness of an individualistic concept of reputation is, however, problematic, for example, because trustworthiness, risk, and reputation are often loosely coupled (e.g. Meyerson, Weick, and Kramer, 1996). The individualistic concept of reputation, of course, also makes major theoretical assumptions about the nature of individuals, social activity, 'information', and human cognition. The question on incentives and reputation, therefore, is an open one.

10.6.1 The Tulip Flame War

In any social system, reputation, authority, and legitimation are products of history, and abstract definitions of them easily fail on a closer study. Reputation is defined within the community in question and the criteria it uses in managing reputation change as the community evolves. There are no 'universal reputations' as reputation forms and is reproduced in ongoing narratives and within a complex ecology of actors whose acts gain them meaning as episodes in a history (Czarniawska, 1997; MacIntyre, 1981). The only way to learn the rules of reputation building is to become engaged in the community discourse. Breaking the rules, in turn, can lead to excommunication.

All social interaction creates conflicts. To alleviate them, communities develop norms and value systems. Durkheim (1933) argued that the essence of modern capitalism was that the increasing importance of division of labour had eroded the basis for communally shared values. In the modernization of the society, insults against communally shared values, in other words crimes against the community—those that were encoded in religion and penal code—were increasingly being replaced by conflicts between parties who regulated their interactions with contracts. This was reflected in the increasing prominence of civil, commercial, procedural, administrative, and constitutional laws in modern society.

In this sense, electronic communities resemble more medieval villages than modern cities. Crimes in an electronic community are crimes against the community. Whereas in modern society punishment is typically based on restitutive sanctions which try to return things as they were before norms were broken, in electronic communities sanctions are often repressive. As Reid (1999: 118) has noted, punishment in electronic communities often shows a return to the medieval, making value conflicts highly visible spectacles that focus on the sins of individuals.[46]

Open source communities, of course, have much less conflict than many other communities. As virtual communities live on the net, their value systems are transparent. Potential community members can easily check if the community adheres to acceptable values. The effectiveness of this self-selection process is clearly indicated, for example, by the fact that over 99 per cent of the people mentioned in the Linux credits file are male.[47]

Sometimes, however, conflicts arise from within the existing community. A recent email exchange in the linux-kernel mailing list gives an example of this process. The weekly Kernel Traffic linux-kernel mailing list summary called this episode 'Tulip driver developer flame war'. There were 71 mailings around the topic between 13 and 20 March 2000. The main issue was the style of development by Donald,[48] one of the people mentioned in the credits file. This exchange of messages shows in a concrete form how behavioural norms are discussed and articulated in the Linux development community. It also shows how technological architecture, development practices, open communication, and reputations interact, and how the Linux community implements reflective practices that maintain the community and facilitate social learning. The Kernel Traffic editors summarized some of the discussion:[49]

In the course of argument, Donald said to Jeff, 'you didn't understand the task you were taking on when you decided to take over maintaining the Ethernet drivers. It took years to write the

---

[46] This is also one important reason why I think Himanen's (2001) proposal that the hacker ethic could be generalized, for example, for development of education is problematic. Instead of being a new model of community-based development, Himanen's hacker ethic has close similarities with the medieval social model, where conflicts between villages were often solved by brute force, or with the ethic of modern fundamentalist religious movements. An attempt to develop an ethically more sophisticated version of the hacker ethic could be based on Habermas's (1993) discourse ethic. The norms that could take into account the different values systems of different communities, in turn, could be developed by studying the theory of conflicts and justice (e.g. Rawls, 1999).

[47] As Kendall (1999: 63–4) has noted, in electronic communities the first entrants often set the norms for the community.

[48] I have edited the citations related to this episode so that they include only the first names of the persons involved.

[49] This is an edited version of the summary in Kernel Traffic #60, 27 Mar. 2000, http://kt.linuxcare.com/.

driver set—it's not something you can just pick up in a few months. And expecting me to now fix or maintain your hacked up code branch is just completely unreasonable.' Jeff replied with venom:

> No one expects anything from you and has not for a long time. If you wanted to actually WORK on the drivers, rather than just complain, then I'm sure many people including myself would find that work very valuable . . .

Elsewhere, Jeff went on, 'Donald, I, and others all seem to agree that having his drivers and the kernel drivers diverge is a poor situation. However, while Donald continues closed source development with periodic code drops, and does not work with other kernel developers when creating infrastructure, I do not see a resolution to the situation any time soon.' David replied angrily, 'Please explain how his code development is closed source? This is totally BS and you know it. All the code is available, all the list discussion is available, and patches and requests are accepted all the time. Quit it. His development is quite open . . . '. Linus replied:

> David, pipe down.
>
> You seem to like the approach Donald has taken. But take it from me, it DOES NOT WORK.
>
> The problem is that maintaining the drivers in their own small universe means that only those people who follow the driver development will ever even test them. . . .
>
> I fixed the tulip driver at least twice to work with the media detection, and sent Donald email about what I had done and why . . . I don't know if my fixes ever actually made it into Donald's version, because after the second time I just stopped bothering trying to re-fix the same thing, and I never updated his driver again.
>
> In contrast, what Jeff and others have done have [sic] been of the type where immediately when a fix is made, it is released. Which means that if there are problems with it, people who follow new kernel releases will know. Immediately. Not in a few months time when the next 'driver release' happens.
>
> This is what Jeff means with 'closed source'. Yes, the sources are there. Yes, they get released every once in a while. But Donald doesn't let people _participate_. He thinks he is the only one who should actually touch the driver, and then he gets very upset when things change and others fix up 'his' drivers to take into account the fact that the interfaces changed. . . .

Jeff also replied to David:

> Donald's development is not open AT ALL. . . . He disappears for many months, creates a design without interfacing with kernel developers, and then appears again with a code drop. It is classic cathedral style of development. Read Eric Raymond's paper on why the bazaar method is far, far superior. . . .

Donald replied to Jeff:

> A quick search of the two very active Tulip mailing lists reveals that you have contributed nothing until this year. Apparently you were not even a subscriber until then, and know nothing about the very open way development has been done. Yet you willingly throw around pejorative phrases like 'cathedral style'—a hot button in this community.
>
> For those not interested what superficially appears to be a kernel power grab, there are issues underlying all of what appears to be a personal conflict.

The Kernel Traffic summarized in more detail Donald's argument that the under-lying questions are about the stability of kernel source code interfaces, testing the drivers in the context of continuously changing kernel releases, and the large and frequent kernel patches that make life difficult for driver developers. Donald further stated that the earlier interfaces were better than the more recent ones, and questioned the viability of the monolithic, single-point kernel source tree. Linus replied:

> You're basically the only one thinking so.
>
> The fairly recent changes in 2.3.x (the so-called 'softnet' changes) are just incredibly more readable and robust than the old crap was that I don't see your point at ALL.
>
> Just about every single network driver out there was SERIOUSLY broken . . . I know, I had fixed many of them. The games the drives played . . . were just incredibly baroque, and had absolutely NOTHING to do with 'clean'.
>
> All of that crap is gone, and it was much overdue. . . .

The Kernel Traffic summary further recorded that Donald was no longer considered to be the owner of the network drivers that he had earlier developed. First Donald lamented the difficult situation he was in because he didn't have sufficient control over the development. Then Linus gave his assessment of the situation. The Kernel Traffic summarized:

Elsewhere in an entirely different subthread, Donald argued:

> I'm in the increasingly untenable position of being expected to maintain drivers for the current and older kernels, but not having any influence over the new development exactly because of that backwards compatibility. It's no fun being responsible for just the old versions, especially after I did years of unpaid development work.
>
> There were many interface changes added incrementally in the 2.3 kernels. Some were added without consideration of, or even in opposition to, cross-version compatibility. And few of those interface changes were designed, as opposed to just hacked in. When I proposed a new PCI detection interface I wrote a skeleton driver, converted several of my drivers, demonstrated that it worked with several hardware classes and wrote a usage guide. But the few day hack was added because the patches were incremental (even if misdesigned and broken).

Linus replied:

> Donald, that's not true, and you know it.
>
> Neither I nor anybody else has expected you to maintain the drivers for quite a long time now–you just didn't seem to have the interest, and a lot of people have acknowledged that. That is why there ARE new maintainers for things like tulip and eepro100, whether you like it or not.
>
> You did not lose influence of the drivers because you want to maintain backwards compatibility. You lost influence over the drivers simply because you never bothered to send in your changes. Don't start blaming anybody else.

As this brief outline of the driver developer flame war shows, the open source model has conflicts, and reputation and authority can be gained and lost. As the comments of Linus Torvalds reveal, breaking of expectations can lead to neglect of contributions, thus effectively destroying the possibilities of gaining reputation within the community. When the reputation has decreased enough, it becomes easy for someone to start parallel development. Eventually this can lead to explicit transfer of 'ownership' rights.

The question is, however, also about the locus of control. Donald, as a driver developer, prefers the kernel to stay stable so that he can more easily develop his software. Linus indicates that in the Linux community, the kernel is the central artefact, and driver developers should adjust to the requirements of kernel development. Donald's position is therefore rather similar in relation to the kernel as kernel developers' position is to the GNU c-compiler. As was noted above, the kernel developers argue that the compiler version needs to be held constant to effectively debug problems in the new kernel releases. The big difference, of course, is that the kernel developer community has to explicitly integrate the driver developer's contributions to the overall system to make them useful. The kernel community can simply say that it does not like the idea.

By following such discussions, novice developers can learn how open source development is interpreted in practice, and what the taboos are that should not be broken. As can be seen from the example above, the open source model implies a code of conduct, which is supported by a socialization process that also occurs in the open source mode. One reason for the effectiveness of the open source development model is that negotiation of social practices and the development of reputations can be observed by the global community in real-time.

11

Concluding Remarks

11.1 LINUX AS MODERN ECONOMY

Since Schumpeter, economists have defined innovation as something that has a direct economic impact. According to this view, inventions and technological change become innovation when they change production functions that relate economic inputs to economic outputs. Much research and many policy recommendations have been based on this economic view of innovation. The historical examples discussed in the previous chapters show, however, that there can be innovative technological change that is not captured by this definition of innovation. For example, Linux, as it exists today, obviously has potentially important implications for the software industry and for the rest of the economy. Yet, it is difficult to argue that economic rationality would explain the evolution of Linux or, indeed, that economic interests would have played a crucial role in its history. Linux developers operate in a field of social interaction that is created as a result of this interaction. As this field is constantly reorganized and expanded, 'economic structure' cannot emerge easily within it. The evolution of Linux, therefore, cannot be described as 'maximization' of any given utility. Strictly speaking, Linux, therefore, is no more based on an 'economy' of gifts than it is based on an economy of monetary transactions. The economics of innovation requires a different—expansive—form of economy. The economics of innovation requires a concept and theory of value that is grounded outside the economy itself.

One way to approach this fundamental challenge is to redefine the concept of value so that it is linked to social practices. Here the analysis of MacIntyre (1981) is particularly relevant. MacIntyre argued that we should distinguish internal and external values. Internal values are the principles that organize the practice and make the activities of the community members part of a meaningful history. For example, for a football team 'a good game' may be a value in a similar way to that in which a 'beautiful code' can be a value for Linux developers. Internal values, then, are the ends-in-themselves for the community in question, and provide the 'virtues' of the community. External values, in contrast, are instrumental means for other ends. Software, for example, can be developed because collaborative development is fun and because it feels good to be able to develop one's competencies as a programmer. These values are internal to the practice in question and they don't make sense outside it. It is also possible to hack the Linux code simply to be able to get a nice sports car or a university degree. From the perspective of the Linux development practice, such values are external.

In practice, internal and external values are difficult to separate. Conceptually, however, they are independent. External values generate the domain of economic transactions. The basic theoretical assumption of modern economic theory is that there are no local currencies or values. In this view, all values are reflected on external values that can be made universally comparable using the market mechanism. The cost of this concept of value is that the social foundation of internal values disappears and the logic of productive communities becomes encrypted and hidden inside a black box which mysteriously emanates useful products. The source of the problem, of course, is in the fact that internal values were thought to be non-existent simply because they did not exist in the abstract world of economic transactions.

The Linux community is a particularly interesting technology-producing community as it lacks the modern system of external values.[1] Until recent years, it has produced only internal values. The external use of the produced results has occurred as an unintended side effect. Indeed, some critics have argued that the Linux community only cares about its own needs and fun and does not, for example, document the system in a way that would allow people in developing countries to utilize the potential of Linux. In this regard, of course, the modern economy and the Linux community are both uninterested in the common good. The modern economy, however, is fundamentally production 'for others'—a labour society, to use Arendt's (1998) term. The difference is that if consumers in developing countries had money, the modern economy would produce external values for them. In the Linux community, the common good is not interesting because it is external to the community and only internal values count. In the labour society, common good does not count because only external values count. In the modern economy, the assumption was that external values are fully expressed using money as a universal signalling mechanism.

Although internal community values are invisible in the modern economy, they often drive development of competences and technology. For example, open source communities are in many ways similar to amateur radio communities. Amateur radio licences explicitly prohibited commercial activities and the amateurs were repeatedly pushed into useless and unusable parts of the radio spectrum as soon as their innovations had prepared the ground—or space—for commercial

[1] As Paul Duguid has emphasized, open source communities of course rely on resources produced by economic actors. There would be no Linux without Intel's microprocessors, university computer networks and file servers, or slack resources in business firms. In this regard, open source communities are economic free-riders. One reason why open source communities are able to live according to their internal values is that the Internet both lowers development costs and makes it possible to distribute them among a large population of developers. As a result, the costs can become invisible to existing accounting systems.

or 'professional' use. Radio amateurs shared and developed technologies and competencies that made, for example, mobile phones possible.

Similarly, electronic bulletin board networks prepared the way for the Internet in the 1980s.[2] Many of the most skilled computer programmers of the 1990s acquired their skills by packing animations and music in 'demos' that simply showed off the capabilities of their programmer or programmer team. Before the breakthrough of the Internet in the early 1990s, these demos were shipped around the globe on magnetic disks like viruses. This all happened despite the fact that collaboration and communication were not particularly easy.[3]

Linux, amateur radio, bulletin board networks, and the computer demo scene provide historical examples that show that innovation does not determine economic change and that economic rationality does not determine technological change. Innovation and economy are only loosely coupled. To describe the linkages between technology and economy, the modern economy itself has to be explained as one of the sophisticated technologies of modern society.[4]

Schumpeter, himself, adopted such a sociologically oriented view on economy. He maintained that it is impossible to separate the processes of technological development and the system of capitalism, arguing that they were essentially one and the same thing (Schumpeter, 1975: 110). Capitalism, therefore, was not a way to organize economic transactions in a system with private ownership, but a specific institutional, cultural, and social form. In itself, capitalism is based on rationality that generates the cost-profit calculus and turns money into a unit of account. By crystallizing this rationality and representing it numerically, cost-profit calculus

---

[2] For example, in 1986 there were over 80 electronic bulletin board systems in Finland. Hackers were combining radio technology, computers, and computer networks (Tuomi, 1987). Several of the active Finnish hackers of the 1980s are now globally known Internet experts and entrepreneurs.

[3] Raymond's (1999: 99) argument that the open source development model relies on the abundance of resources seems to assume that resource use and ownership are tightly connected. As Duguid argues, open source developers often use resources owned by others. The history of amateur radio, however, also shows that resource scarcity does not stop innovation and collaboration. On the contrary, sometimes it generates constraints that become challenges for the innovators. For example, in 1964, after two years of intensive experimentation, the first successful two-way radio connection was made by bouncing radio waves from the moon (Janhunen, 1964). This two-person earth–moon–earth network, between William Conkelin, Long Beach, California, and Lenna Suominen, Nakkila, Finland, shows that breakthroughs are sometimes generated with great persistence and effort, when experts argue that the goal is impossible, and without any perceivable economic spillovers.

[4] Simmel's (1990) observation was that money is the most perfect tool. One could argue that modern economic theory has focused on one particular social interface and translation mechanism, the social technology of money, taking the characteristics of this technology for granted. A sociological and cultural starting point would lead us from the closed self-referential system of economic utility relations to discuss economic systems as forms of socio-technical systems where other tools and technologies are also relevant. Such a theory might be based on sociocultural activity theory (Leont'ev, 1978), which inherently combines technology, social division of labour, and human cognition, and which could be expanded to include accumulation and creation of different types of resources, for example, technologies, conceptual systems, and knowledge. Indeed, Engeström (1987) has developed a version of activity theory where exchange, division of labour, productive action, and technology are parts of the same structure of human activity. Engeström's focus, however, was on the theory of learning.

creates and promotes the modern enterprise. Economic rationality, however, does not stop there:

And thus defined and quantified for the economic sector, this type of logic or attitude or method then starts upon its conqueror's career subjugating—rationalizing—man's tools and philosophies, his medical practice, his picture of the cosmos, his outlook of life, everything in fact including his concepts of beauty and justice and his spiritual ambitions.   (Schumpeter, 1975: 123–4)

Schumpeter pointed out that the capitalistic enterprise is founded on institutions that necessarily become decomposed as economic rationality becomes increasingly dominant. The historical patterns of capitalistic consumption and accumulation, for example, were based on a family structure where individual interests could become intergenerational. The traditional bourgeois self-interest of entrepreneurs and capitalists was historically never very individualistic and both the rhetoric and action were impregnated with the idea that the entrepreneur was acting on behalf of future generations. Family and family home used to be the mainspring of the typical profit motive (Schumpeter, 1975: 160). As the institution of the bourgeois family increasingly came to be viewed through economic rationality, Schumpeter noted that this foundation of capitalism was about to erode:

As soon as men and women learn the utilitarian lesson and refuse to take for granted the traditional arrangements that their social environment makes for them, as soon as they acquire the habit of weighing the individual advantages and disadvantages of any prospective course of action—or, as we might also put it, as soon as they introduce into their private life a sort of inarticulate system of cost accounting—they cannot fail to become aware of the heavy personal sacrifices that family ties and especially parenthood entail under modern conditions and of the fact that at the same time, excepting the cases of farmers and peasants, children cease to be economic assets.   (Schumpeter, 1975: 157)

Schumpeter pointed out that these sacrifices do not consist only of monetary sacrifices, but comprise an indefinite amount of loss of comfort, of freedom from care, and opportunities to enjoy the increasing variety of attractive alternatives.[5] According to Schumpeter (1975: 158), the questions that economic rationality makes parents ask can be summarized: 'Why should we stunt our ambitions and impoverish our lives in order to be insulted and looked down upon in our old age?'

Free markets enable individuals to make choices based on their preferences. Freedom, however, is a social phenomenon, and it is bought at the expense of others. Individuality itself is a modern invention (Taylor, 1989). Individual choice is possible in the free market when everyone in the market buys the logic of economic rationality. Economic rationality is one of the distinctive achievements of the modern world, but its operation is possible only because it operates in a complex institutional and ideological context.

---

[5] Martin Carnoy (2000) has argued that increasing globalization, competition, and demand for a flexible workforce require new integrative institutions and that they have already transformed families. His description of the changes of family structure is based on recent data, which seems to support Schumpeter's analysis of the consequences of capitalism in this regard.

Linux shows that in real life the assumption of universal economic rationalizing is not necessarily valid. Modernity cannot be reduced to the modern economy, even in the domain of production. The rationality that defines the logic of open source development is a rationality of meaningful social action. It expands the opportunities of human action, instead of optimizing them.

As Schumpeter observed, technological development is linked to social institutions and the values that underlie them. These institutions provide resources for development, but not all development fits easily within their constraints. Why, then, does a complex technological system such as Linux seem to be so successful? Where are its core institutions? Where is its home?

Berman (1982) characterized the modern mentality as a set of culturally shared beliefs. The modern mentality, according to Berman, is composed of strong individuality, belief in self-directed reason, assumption of the individual as the locus of control, commitment to progressive improvement, a generally optimistic outlook, and a strong belief in meritocracy and social mobility. In the Linux community, these beliefs are easy to detect. The proponents of open source often adhere to a very strict rhetoric of meritocracy (e.g. Raymond, 1999; Wayner, 2000). In theory, everyone is judged based on the quality of the results, and seniority comes into play only in those exceptional cases where peers cannot judge quality, or when ownership rules do not work. The great commitment and enthusiasm of Linux developers indicate that the developers believe that their efforts and contributions matter, and that the system as a whole is improved as a result of these efforts. The joy of hacking, in turn, is very much about getting control over a constructed world, and becoming a wizard in such a technological world.

The way the Linux development community lives, indeed, reflects major currents in the modern world. This is also probably one of the reasons why it has been so successful in technological development. The social system of the Linux community is not only aligned with the needs of the community itself. It is also aligned with important components of the broader social system where the development transpires.[6]

The great paradox of Linux is that although its developer community subscribes to the core values of modernity, at the same time its social structure resembles

[6] Borsook (2000) argued that computer hackers are greatly influenced by libertarian ideals, where extreme individualism mixes with belief in technological solutions to individual and social problems. As always, reality is more complex than any of its descriptions, and it is possible to find hackers with different world-views. It is, however, clear that discussions on technological development and the future are often expressions of existing interests, conflicts, and values. The Internet, for example, has often been argued to be a fundamentally important technology because it will bring democracy, freedom, and expanded opportunities for choice, and because it will destroy the old structures of power. Although there may be examples of all these developments in the recent history of the Internet, there is little systematic evidence for such technological determinism. It is also quite clear that, for example, democracy will be understood quite differently in an Internet-enabled world than it has been understood in the various phases of Western political history. The Internet probably will change the processes of opinion formation, which, in turn, will change the institutions of democracy. The exact forms of change, however, remain open for research. The safest bet is that the reality will surprise us.

a medieval village. Social cohesion is based on shared values in the Linux community. The use of the Internet, however, has created a new virtual village of transpatial solidarity, where Durkheim's value-based mechanical solidarity and organic solidarity coexist happily. This is possible because the Internet has made the village large enough so that it can effectively organize the division of labour, without giving up its values.

The system of values in the Linux community is that of modernity and is therefore in many ways compatible with the values of entrepreneurial capitalism. Why, then, has Linux development not been organized using economic transactions? One reason seems to be that, as the developers engage in a joint effort, it has been difficult to atomize social interactions into economic transactions. The economy of Linux is an economy of networks. As long as the developers produce a system with many interdependencies, there is no easy way to 'clear' the transactions or represent them using a one-dimensional measure, such as money.

Instead of clearing social transactions on the spot, open source developers make them reversible. Copyleft and the availability of the source code facilitate this. Because social transactions are more reversible than market transactions—simply because market transactions aim at making transactions historyless—developers can become more interdependent. In other words, they can rely on each other's skills and capabilities more than they could under pure market conditions. This possibility is especially important in creative and non-routine knowledge-based work, such as Linux development. The evolving source code, therefore, provides a platform on top of which social capital can be accumulated.

The culture of hacking is probably the most perfect and frictionless implementation of modernity, and therefore it also produces technological products effectively. There are no deep internal conflicts that would compromise its competence-based efficiency. As long as it builds itself around those technological artefacts that it produces, it is able to avoid many of those conflicts that make similar efficiency difficult in broader social contexts.

It would, indeed, be surprising to see successful technology development projects in communities where the values of modernity were strongly contested. Could we imagine successful collaborative development of a technological artefact in a cultural setting where the developers would believe in unpredictable accidents, irrelevance of one's own interests and decisions, belief in the inevitable deterioration of the developed system, and questionability of the meaning of the whole effort? In this sense, successful technology development requires modern values. A different set of values would require a very different concept of technology and technological progress. But not all technology is created equal. Even within modern high-tech organizations there are important cultural differences that may constrain or facilitate open source development. For example, the open source development model seems to require a culture with low power distance and low uncertainty avoidance. It is well known that the regions of the world differ greatly in this respect. Indeed, it may have some relevance that Finland, the country where the story of Linux started, and the USA, the mother lode of the high-tech gold rush, happen to be countries with the least power distance and uncertainty avoidance (Hofstede, 1991).

The Linux community consists of members who live in many different local cultures. It is therefore an interesting case of a 'global' culture, and it might provide important insights on the mechanism that links regional cultural resources to global technological development. As many economic actors currently try to integrate their activities with open source communities, the analysis of the successes and failures of this 'civilizing' and 'modernizing' process could also provide important insights on the nature of the emerging second modernity.

11.2 THE HIERARCHY OF INNOVATION

Following Schumpeter, several authors have argued that new key technologies generate long waves in the economy.[7] Perez (1985) noted that before productivity of a new key technology can be realized, existing forms of production, organizational structures, banking and credit system, and other social institutions have to change. According to this institutional interpretation of economic long waves, the rigidity of social institutions is the reason why long economic waves are long.

The discussion on institutional change focused on the social basis of macroeconomic growth. It highlighted the fact that the diffusion and impact of new technologies cannot be understood by looking only at the characteristics of artefacts. In this view, technical innovations require complementary social innovations, which often are the main bottlenecks in the adoption of innovations.

Although this view is insightful and useful, it is difficult to connect with the actual use of technologies. This is because conceptually it separates technical and social innovations, implicitly reproducing the distinction between artefacts and their meaning. This is philosophically a bad choice, and it misses the essence of innovative change. Artefacts embed meaningful uses. Meaningful uses, in turn, gain their meaning from social practices. Innovation, therefore, is a social, material, and interpretational complex. Innovation can occur in any of these dimensions, but often it is about simultaneous change in all of them.

One way to describe analytically these different dimensions of innovation can be based on Leont'ev's activity hierarchy. As was noted above, Leont'ev distinguished three levels in activity. The level of activity itself is the level where activity can be understood as meaningful social practice. Activity, in turn, is implemented at the level of actions. This level of action is visible for outside observers, but the meaning of related activity is as impossible to derive from actions as it is to derive the meaning of a sentence by concatenating definitions of words. Actions, finally, are made concrete in specific contexts at the level of operations. For example, we can hear someone beating a drum. Without knowing what the drummer is doing we cannot know if she is trying to frighten game, wolves, or lions; please spirits,

---

[7] As was noted before, the theory of long waves has a long and controversial history (cf. Freeman, Clark, and Soete, 1982; Kleinknecht, 1987; Berry, 1991; Mandel, 1995).

generals, or the court; or earn her living in a jazz band. The drum can be an empty skull, hollow tree, parade snare, or an electronic device connected to a synthesizer.

Innovation can occur in all these levels of human activity. At the level of operations we can talk about technical innovation. Innovation at this level is concerned with tools that enable and constrain operations. At the level of action, activity is decomposed into sequences of goal-oriented actions that implement activity. At this level, therefore, innovation is concerned with problem-solving strategies and creative use of intended and unintended resources.

Change at the level of activity, in turn, has two independent sources. One is related to the internal values of the underlying community of practice. New motives and objectives may emerge and activity can acquire new meaning. Hunting, for example, can be transformed from a food-producing activity into a sport. The meaning of activity can also change because the relations between different communities become reorganized. Such a configurational change cannot be understood simply from the perspective of a single focal community; instead, it occurs at the level of an ecology of practices.

Using the hierarchy of activity, we can therefore climb from the level of technical artefacts all the way to the level of external values, linking technological change to economic change. Instead of simple progression from the lower levels of hierarchy towards the higher levels, the levels mutually construct each other.

Moving down from the higher levels of activity hierarchy, we can see how actions and operations gain their meaning. Moving up, we can see how action, activity, and communities gain their resources and configure their constraints and concrete contexts. The technological development of societies occurs on these multiple levels simultaneously.

Innovative change is construction of meaning. The created meaning can become social when it is externalized in symbolic representations and artefacts. But as the material, mental, and social constraints are generated in a historical process of co-evolution, mind and matter are always mutually constructed. As Nishida argued, we do not simply construct our reality out of thin air but the reality also constructs us. Sometimes we surprise the reality—make real things that are novel. And we ourselves become real in a material, cultural, and social context. And often the reality surprises us.

11.3 THE NEW ECONOMY

One of the consequences of the Internet may be that technology development is increasingly unlinked from local social institutions. In the case of Linux we can see a process where the social system and its institutions are continuously negotiated, on-line, reflecting the problems and opportunities generated in the process. Linux—and other Internet-based innovations—provide examples of socio-technical development that perhaps escape the logic of long waves, and which potentially break long waves into continuous ripples. The innovation process that underlies

Linux development, therefore, could also give a concrete example of what the 'new economy' is about. But, as we saw in the case of Linux, new institutions may be needed to keep the development going. The institutionalized core kernel of Linux, therefore, is an interesting example of a generic process in which social structure becomes embedded in technology. As Lessig (1999) notes, such embedding is becoming increasingly common and creates new policy challenges.[8]

Linux development is based on a complex interplay between social practices and a focal technological artefact. Any single driving force, for example financial rewards, cannot explain Linux development. In this sense, open source development is similar to medieval cathedral building. We do things that our peers appreciate and which are meaningful in the social context we are in. Linux development, therefore, is not a result of any specific economy based on transactions, bartering, or exchange of gifts. Instead, it is better characterized as a practice and a form of social life. The artefact that organizes this form of life emerges as people go on with their lives in ways that are meaningful to them. In this sense, Linux development is totally endogenous: the technological artefact can be seen as a side-effect of the fact that people live and construct their identities in a social world that is organized around this technological artefact, which then becomes the stage for feasts and congregations, and the centre of community life (Branner, 1961).

New technologies are always appropriated by integrating them into social practice. Indeed, it was argued above that innovation occurs only when social practice changes. Often such change results from appropriation of a new tool which reorganizes the practices of a community. The key to innovation, therefore, is in those social communication and learning processes that underlie change in social practices.

Social practices, however, are interlinked in the ecology of communities. It is not always possible to change social practice without breaking those translation processes that make a community a resource for other actors. Change is difficult especially when the same resource is used by several actors. As the evolution of Linux shows, one way to solve this problem is to sediment resources, institutionalize practices, and stop innovating.

The history of Linux, however, also shows that effective resource translation mechanisms can lead to rapid growth. In software development the problem of managing interfaces between modules has led to relatively standardized ways of building modules and using interfaces. This, in turn, means that modules can easily be added to the system. Furthermore, these standardized translation

---

[8] DeLong and Froomkin (2000) have also argued that informational products create major policy challenges as the basic assumptions underlying the efficiency of the market mechanism are becoming increasingly invalid. One of the drivers of the quest for a new theory of political economy is the fact that 'products' are becoming increasingly long-term relationships between producers and consumers. In Victor and Boynton's (1998) terms, products require continuous co-configuration of production and use. Adam Smith's invisible hand, therefore, is attached to an increasingly blind body that is ignorant of costs and benefits of entering an economic relationship. The theory of free markets, obviously, has great difficulty in handling such situations. Linux development is a particularly clear example of such long-term relationships of co-configuration.

mechanisms mean that modules can be relatively easily used by different actors even when the modules change. In the history of Linux, we can see this magic of black-boxing in operation.

Linux is therefore in many ways open to combinatorial innovation. Standardized interfaces and translation processes generate smooth module boundaries and facilitate rapid recombination. Sometimes the source code itself can be reused, but more importantly the learning that is represented in the source code can be reapplied in different contexts without major problems. As a result, the various communities that develop the different parts of the Linux kernel become very mobile. In this way, the solution to the problem of translation leads to an ecology of communities that can readily reconfigure its resources.

When money is used to coordinate social interactions, new combinations can be created with little friction. Under these circumstances, however, combination also easily destroys social capital. Economic rationality reduces history into a sequence of 'choices' and atomizes the value of social transactions. From these 'atomized' transactions, networks can be built only by linking sequences of essentially independent events. Open source and standardized interfaces, in contrast, enable a system where collective interactions can be managed. This is a fundamental difference between a traditional economic market and open source collaboration. This is also the reason why already in its genetic origin open source development is a networked form of collective production. A developer sub-community can simultaneously translate its resources for several user communities. This also means that several sub-communities can become involved in those innovation and development processes that produce new technology. Whereas, in the theoretical model of a traditional economy, only money talks, the open source mode of development is an inherently polyphonic system of development.

In Linux development, the Schumpeterian creative destruction destroys pieces of code, but competence and experience are reorganized with little waste. In this sense, one could argue that the Linux development model and the Silicon Valley innovation model (Kenney, 2000) have similar characteristics. The main difference, of course, is that Silicon Valley has a venture capital driven entrepreneurial culture, whereas the economic sphere has been relatively invisible in the Linux development. As was noted above, free market capitalism and open source development are fundamentally based on the same modern values. The difference is that market capitalism makes economic rationality one of its core values, whereas open source developers may regard it as less central. This tension is actively being managed by the Open Source Initiative. Indeed, the Open Source Initiative can be seen as one more institutional structure that pops up in the evolution of Linux to repair social damage that is created when these two relatively similar cultures collide and create conflicts in the developer community.

As the analysis of the evolution of Linux shows, rapid growth requires that the core is institutionalized and that some of the translation processes are taken for granted. In this model, innovation happens in periphery. It is interesting that such peripheries are conventionally described as frontiers. We could, however, ask whether—and in what sense—progress results from moving the boundaries of periphery, or whether this is simply one strategy to reduce change in the core.

Today, we live in an age of individualism. Technological and social change is fast, and traditions are mixed and matched according to individual preferences. In the public discourse, social utopias have been pushed to the periphery and the focal vision reveals a world where everyone has the responsibility and right to value and cash their life options. Yet, perhaps it is also clear that to make meaningful choices, somehow meaning has to be included in the equation. Meaning is grounded in history and social interaction. Individuality itself depends on shared history, and we become individuals only in relation to others. Linux and the Internet, therefore, were not produced because they were solutions to a well-defined problem or because their development would have been economically beneficial. Instead, they were produced as side-effects of meaningful social interaction. On the Internet, much of this interaction becomes documented and available for further study. This is also the reason why Internet-related innovations make visible those social processes that underlie innovation and technology development. In the history of Linux, innovation can be seen in its bare form, without the legitimizing veil of economic rationality. In the history of the Internet and packet-switching networks, we have to follow a more complex route through layers of economic rhetoric and rationalizing. Eventually, however, the social core of innovation becomes so obviously apparent that it is impossible to neglect any more.

11.4 THE ROAD AHEAD

Technological innovation is more than the production of improved functionality. In open source projects it is easy to see that striving for the common good is one of the reasons why open source developers commit themselves to a development project. Although the common good is evaluated based on the internal values of the community, even a small contribution can become important when it becomes part of a bigger system. Most developers are not interested in positive economic returns. They are interested in a great story, and the possibility to be part of it. As long as people can make history, there is no end to history.

Innovation, therefore, has its deep roots in the processes of individuation, socialization, and meaning construction. We use language, signs, and tools, and integrate them in our thinking and action. In this sense, human beings are technological beings. Fundamentally, technological change, therefore, relates to questions concerning the way we exist in the world. As technologies and technological change become increasingly visible in our everyday life, the foundations of technology also will be increasingly in our focus.

Above we saw many examples of those social processes that underlie innovation. Due to the increasingly knowledge-intensive and networked nature of innovation processes, the social dimension of innovation is becoming increasingly difficult to neglect. The heroic model of innovation is being replaced by a social model of innovation, and the various communities that provide resources for innovative change will certainly be studied with great interest in the future. Research on the politics of innovative ecologies will emerge as one important source of future innovation policies. Eventually, however, we also have to ask what our relation is to technology and technological change. To answer such questions, we have to develop theoretical concepts that allow us to simultaneously discuss meaning creation, human cognition, social activity, and ethics. Today, attempts for such theoretical discussion may seem philosophical and remote from the practical work of innovators. Partly the reason is that we don't really have very much empirical research that could provide concrete starting points for such discussions.

Although it is certainly difficult to predict the impact of new technologies, in the future there will be increasing demand for processes that integrate technology development with policy. This is simply because we already know that technological development does not imply universal progress. Whereas technology policy traditionally has been closely linked with national competition policies, in the future the impact of technologies and policies will be increasingly global. Perhaps, as Ulrich Beck (1994), for example, has proposed, we need new institutions that negotiate and integrate the different interests and voices that interpret the meaning of technology.

Internet-related innovations are interesting as they show how multiple interests can be combined so that everyone is better off at the end of the day. They are also interesting because, due to the networked nature of these innovations, the processes that produced them are networked. In Internet-related innovations it is difficult to neglect the fact that innovation is a social process. It is, however, not clear whether the innovations discussed in the previous chapters fully represent all the various types of innovative activity. We have focused mainly on a small set of innovations that obviously have great impact, but which are also in many ways special. One may wonder whether the discussion in the previous chapters can readily be extended also to those innovations where software and communication technology play only a minor role. One may also wonder what all the previous discussion means for organizing product creation in firms.

The obvious answer is that there is much more to be said. Many of the conclusions presented above are tentative and can be contested. Although for some readers some of the claims made in the previous chapters may seem radical, I believe in incremental change and small contributions. As the case of Linux shows, incremental modifications may have a big effect when they focus on critical points in the overall system. For me, one such critical point is that there have been few empirical cases that would enable us to open the black box of innovation, and allow us to further study those social meaning creation processes that make technology what it is. This does not mean that economic considerations would be irrelevant. It simply means that economic considerations are not enough. Innovation is not only about better functionality or about economic rationality. These are parts of a more complex picture.

References

Abbate, J. (1999). *Inventing the Internet*. Cambridge, Mass.: The MIT Press.

Adler, P. S. (1988). Managing flexible automation. *California Management Review*, 30 (3).

Akscyn, R. M., McCracken, D. L., and Yoder, E. A. (1988). KMS: a distributed hypermedia system for managing knowledge in organizations. *Communications of the ACM* 31 (7): 820–35.

Allen, T. J. (1977). *Managing the Flow of Technology*. Cambridge, Mass.: The MIT Press.

—— and Cohen, S. I. (1969). Information flow in research and development laboratories. *Administrative Science Quarterly*, 14: 12–19.

Alpers, S. (1988). *Rembrandt's Enterprise: The Studio and the Market*. Chicago: The University of Chicago Press.

Anderson, P., and Tushman, M. L. (1990). Technological discontinuities and dominant designs. *Administrative Science Quarterly*, 35: 604–33.

Arendt, H. (1998). *The Human Condition*. Chicago: The University of Chicago Press.

Armstrong, M. (1998). Kernel architecture. http://se.math.uwaterloo.ca/~mnarmstr/report2.

Astrahan, M. M., and Jacobs, J. F. (1983). History of the design of the SAGE computer—the AN/FSQ-7. *IEEE Annals of the History of Computing*, 5 (4): 343–4.

Axel, E. (1997). One developmental line in European Activity Theories. In M. Cole, Y. Engeström, and O. Vasquez (eds.), *Mind, Culture, and Activity: Seminal Papers from the Laboratory of Comparative Human Cognition* (pp. 128–46). Cambridge: Cambridge University Press.

Axtell, G. S. (1991). Comparative dialectics: Nishida Kitaro's logic of place and Western dialectical thought. *Philosophy East and West*, 41 (2): 163–84.

Bakhtin, M. (1987). *Speech Genres and Other Late Essays*. Austin, Tex.: University of Texas Press.

Baran, P. (1964). On Distributed Communications: vols. I–XI. The RAND Corporation. http://www.rand.org/publications/RM.

Barnes, S. E. (1997). Douglas Carl Engelbart: developing the underlying concepts for contemporary computing. *IEEE Annals of the History of Computing*, 19 (3): 16–26.

Beck, U., Giddens, A., and Lash, S. (1994). *Reflexive Modernization: Politics, Tradition and Aesthetics in the Modern Social Order*. Cambridge: Polity Press.

Bergson, H. (1977). *The Two Sources of Morality and Religion*. Notre Dame: University of Notre Dame Press.

—— (1983). *Creative Evolution* (1st edn. 1907). Lanham, Md.: University Press of America.

Berman, M. (1982). *All That Is Solid Melts into Air*. New York: Simon and Schuster.

Berners-Lee, T. (1990). Information Management: A Proposal. http://www.w3.org/History/1989/proposal.html.

Berners-Lee, T., and Cailliau, R. (1990). WorldWideWeb: Proposal for a HyperText Project. http://www.w3.org/Proposal.html.

—— and Fischetti, M. (1999). *Weaving the Web: The Original Design and Ultimate Destiny of the World Wide Web by its Inventor*. San Francisco: HarperCollins.

Berry, B. J. L. (1991). *Long-Wave Rhythms in Economic Development and Political Behavior*. Baltimore: The Johns Hopkins University Press.

Bezroukov, N. (1999). A second look at the Cathedral and the Bazaar. *First Monday*, 4 (12). http://firstmonday.org/issues/issue4_12/bezroukov/.

Bhattacharya, S., Krishnan, V., and Mahajan, V. (1998). Managing new product definition in highly dynamic environments. *Management Science*, 44 (11): S50–S64.

Bijker, W. E. (1987). The social construction of Bakelite: toward a theory of invention. In W. E. Bijker, T. P. Hughes, and T. Pinch (eds.), *The Social Construction of Technological Systems: New Directions in the Sociology and History of Technology* (pp. 159–87). Cambridge, Mass.: The MIT Press.

—— (1997). *Of Bicycles, Bakelites, and Bulbs: Toward a Theory of Sociotechnical Change*. Cambridge, Mass.: The MIT Press.

—— and Law, J. (1992). *Shaping Technology / Building Society: Studies in Sociotechnical Change*. Cambridge, Mass.: The MIT Press.

Blumenberg, H. (1985). *Work on Myth*. Cambridge, Mass.: The MIT Press.

Borsook, P. (2000). *Cyberselfish: A Critical Romp Through the Terribly Libertarian Culture of High-Tech*. New York: PublicAffairs.

Bowker, G. C., and Star, S. L. (1999). *Sorting Things Out: Classification and its Consequences*. Cambridge, Mass.: The MIT Press.

Bowman, I., Siddiqi, S., and Tanuan, M. C. (1998). Concrete Architecture of the Linux Kernel. University of Waterloo, Department of Computer Science. http://plg.uwaterloo.ca/~itbowman/CS746G/a2/.

Bradner, S. (1999). The Internet Engineering Task Force. In C. DiBona, S. Ockman, and M. Stone (eds.), *Open Sources: Voices from the Open Source Revolution* (pp. 47–52). Sebastopol, Calif.: O'Reilly & Associates, Inc.

Branner, R. (1961). *Gothic Architecture*. New York: Braziller.

Brooks, F. P. (1995). *The Mythical Man-Month: Essays on Software Engineering*. Reading, Mass.: Addison-Wesley.

Brown, J. S., Collins, A., and Duguid, P. (1989). Situated cognition and the culture of learning. *Educational Researcher*, 18: 32–42.

—— and Duguid, P. (1991). Organizational learning and communities of practice: toward a unified view of working, learning, and innovation. *Organization Science*, 2: 40–57.

—— and —— (2000). *The Social Life of Information*. Boston, Mass.: Harvard Business School Press.

—— and —— (2001). Knowledge and organization: a social-practice perspective. *Organization Science*, 12 (2): 198–213.

Brown, S. L., and Eisenhardt, K. M. (1995). Product development: past research, present findings, and future directions. *Academy of Management Review*, 20 (2): 343–78.

Buber, M. (2000). *I and Thou*. New York: Free Press.

Bush, V. (1945). As We May Think. *Atlantic Monthly*, 176 (1): 101–8. http://www.theatlantic.com/unbound/flashbks/computer/bushf.htm.

Callon, M. (1987). Society in the making: the study of technology as a tool for sociological analysis. In W. E. Bijker, T. P. Hughes, and T. J. Pinch (eds.), *The Social Construction of Technological Systems: New Directions in the Sociology and History of Technology* (pp. 83–103). Cambridge, Mass.: The MIT Press.

—— (1992). The dynamics of techno-economic networks. In R. Coombs, P. Saviotti, and V. Walsh (eds.), *Technological Change and Company Strategies: Economic and Sociological Perspectives*. London: Academic Press.

—— Law, J., and Rip, A. (1986). *Mapping the Dynamics of Science and Technology: Sociology of Science in the Real World*. Houndmills, Basingstoke: The Macmillan Press Ltd.

Carnoy, M. (2000). *Sustaining the New Economy: Work, Family, and Community in the Information Age*. Cambridge, Mass.: Harvard University Press.

Carter, R. E. (1997). *The Nothingness Beyond God: An Introduction to the Philosophy of Nishida Kitaro*. St. Paul, Minn.: Paragon House.

Castells, M. (1996). *The Information Age: Economy, Society and Culture: Volume I: The Rise of the Network Society*. Cambridge, Mass.: Blackwell.

—— (2001). *The Internet Galaxy: Reflections on the Internet, Business, and Society*. Oxford: Oxford University Press.

Cerf, V. (1999). RFCs—The Great Conversation. RFC 2555: ftp://ftp.isi.edu/in-notes/rfc2555.txt.

Ceruzzi, P. E. (1998). *A History of Modern Computing*. Cambridge, Mass.: The MIT Press.

Christensen, C. M. (1997). *The Innovator's Dilemma*. Boston: Harvard Business School Press.

Cohen, W., and Levinthal, D. A. (1989). Innovation and learning: the two faces of R&D. *The Economic Journal*, 99 (September): 569–96.

Cohen, W. M., and Levinthal, D. A. (1990). Absorptive capacity: a new perspective on learning and innovation. *Administrative Science Quarterly*, 35: 128–52.

Cole, M. (1986). *Culture in Mind*. Cambridge, Mass.: Harvard University Press.

—— (1996). *Cultural Psychology: A Once and Future Discipline*. Cambridge, Mass.: The Belknap Press of Harvard University Press.

Cole, S. (1970). Professional standing and the reception of scientific discoveries. *American Journal of Sociology*, 76 (2): 286–306.

Collins, H. M. (1975). The seven sexes: a study in the sociology of a phenomenon, or the replication of experiments in physics. *Sociology*, 9: 205–24.

—— (1987). Expert systems and the science of knowledge. In W. E. Bijker, T. P. Hughes, and T. Pinch (eds.), *The Social Construction of Technological Systems: New Dimensions in the Sociology and History of Technology* (pp. 329–48). Cambridge, Mass.: The MIT Press.

Conklin, J. (1987). Hypertext: an introduction and survey. *IEEE Computer*, 2 (9): 17–41.

Constant, E. W. (1980). *The Origins of the Turbojet Revolution*. Baltimore: Johns Hopkins University Press.

—— (1984). Communities and hierarchies: structure in the practice of science and technology. In R. Laudan (ed.), *The Nature of Technological Knowledge: Are Models of Scientific Change Relevant?* (pp. 27–46). Dordrecht: Reidel.

Constant, E. W. (1987). The social locus of technological practice: community, system, or organization? In W. E. Bijker, T. P. Hughes, and T. J. Pinch (eds.), *The Social Construction of Technological Systems: New Directions in the Sociology and History of Technology* (pp. 223–42). Cambridge, Mass.: The MIT Press.

Cooper, R. G., and Kleinschmidt, E. J. (1991). New product processes at leading industrial firms. *Industrial Marketing Management*, 20: 137–47.

Corbató, F., Merwin-Dagget, M., and Daley, R. C. (1961). An experimental time-sharing system. Proc. Spring Joint Computer Conference, AFIPS. Excerpts reprinted in *IEEE Annals of the History of Computing*, 14 (1): 31–2.

Crocker, S. (1969). Documentation Conventions. RFC 3: ftp://ftp.isi.edu/in-notes/rfc003.txt.

—— (1987). The Origins of RFCs. RFC 1000: ftp://ftp.isi.edu/in-notes/rfc1000.txt.

—— (1999). The First Pebble: Publication of RFC 1. RFC 2555: ftp://ftp.isi.edu/in-notes/rfc2555.txt.

Csikszentmihalyi, M. (1990). *Flow: The Psychology of Optimal Experience*. New York: HarperCollins.

—— (1996). *Creativity: Flow and the Psychology of Discovery and Invention*. New York: HarperCollins.

—— and Rochberg-Halton, E. (1981). *The Meaning of Things: Domestic Symbols and the Self*. Cambridge: Cambridge University Press.

Czarniawska, B. (1997). *Narrating the Organization: Dramas of Institutional Identity*. Chicago: The University of Chicago Press.

Daniels, H. (1996). *An Introduction to Vygotsky*. London: Routledge.

David, E. E., Jr, and Fano, R. M. (1965). Some Thoughts About the Social Implications of Accessible Computing. Excerpts reprinted in *IEEE Annals of the History of Computing*, 14 (2): 36–9.

Davies, D. W. (1965). Proposal for the Development of a National Communication Service for On-Line Data Processing. National Physics Laboratory, 15 December. In National Archive for the History of Computing. http://www.cs.utexas.edu/users/chris/nph/DaviesLetter.html.

—— (1970). C.C.I.T.T. Meeting at Geneva, November 23 to 27. http:// www.csu.utexas.edu/users/chris/nph/DaviesLetter.html.

—— (1982). Historical note on the early development of packet switching. http:// www.cs.utexas.edu/users/chris/nph/DaviesLetter.html.

Davis, R., Samuelson, P., Kapor, M., and Reichman, J. (1996). A new view of intellectual property and software. *Communications of the ACM*, 39 (3): 21–30.

de Certeau, M. (1988). *The Practice of Everyday Life*. Berkeley: University of California Press.

de Rosnay, J. (1979). *The Macroscope: A New World Scientific System*. New York: Harper and Row.

DeLong, J. B., and Froomkin, A. M. (2000). Speculative microeconomics for tomorrow's economy. *First Monday*, 5 (2). http://firstmonday.org/issues/issue5_2/delong/.

Dewey, J. (1991). *How We Think*. Buffalo: Prometheus Books.

DiBona, C., Ockman, S., and Stone, M. (1999). *Open Sources: Voices from the Open Source Revolution*. Sebastopol, Calif.: O'Reilly & Associates, Inc.

Ditlea, S. (1998). Ted Nelson's big step. *Technology Review*, 101 (5): 44–8.

Dosi, G. (1982). Technical paradigms and technological trajectories—a suggested interpretation of the determinants and directions of technological change. *Research Policy*, 11 (3): 147–62.

—— Freeman, C., Nelson, R. R., Silverberg, G., and Soete, L. (1988). *Technical Change and Economic Theory*. London: Pinter Publishers.

Dougherty, D. (1992). A practice-centered model of organizational renewal through product innovation. *Strategic Management Journal*, (Summer): 1377–92.

Douglas, M. (1987). *How Institutions Think*. London: Routledge and Kegan Paul.

—— (1996). *Thought Styles: Critical Essays on Good Taste*. London: Sage Publications.

Doz, Y., and Hamel, G. (1997). The use of alliances in implementing technology strategies. In M. L. Tushman and P. Anderson (eds.), *Managing Strategic Innovation and Change: A Collection of Readings* (pp. 556–80). Oxford: Oxford University Press.

Durkheim, E. (1933). *Division of Labor in Society*. New York: The Free Press.

DuVal Smith, A. (1999). Problems of conflict management in virtual communities. In M. A. Smith and P. Kollock (eds.), *Communities in Cyberspace* (pp. 134–63). London: Routledge.

Eigen, M., and Schuster, P. (1979). *The Hypercycle*. Berlin: Springer.

Eliade, M. (1991). *The Myth of the Eternal Return: or, Cosmos and History*. Princeton: Princeton University Press.

Engelbart, D. C. (1963). A conceptual framework for the augmentation of man's intellect. In P. W. Howerton and D. C. Weeks (eds.), *Vistas in Information Handling* (pp. 1–29). Washington, DC: Spartan Books.

Engeström, Y. (1987). *Learning by Expanding: An Activity Theoretical Approach to Developmental Work Research*. Helsinki: Orienta Konsultit.

—— (1999). Innovative learning in work teams: analyzing cycles of knowledge creation in practice. In Y. Engeström, R. Miettinen, and R.-L. Punamäki (eds.), *Perspectives in Activity Theory* (pp. 377–404). Cambridge: Cambridge University Press.

——Miettinen, R., and Punamäki, R.-L. (1999). *Perspectives in Activity Theory*. Cambridge: Cambridge University Press.

Fano, R. M. (1963). Proposal for a Research and Development Program on Computer Systems. Excerpts published in *IEEE Annals of the History of Computing*, 14 (2): 9–10.

—— (1967). The computer utility and the community. IEEE International Convention Record, 30–34 Excerpts reprinted in *IEEE Annals of the History of Computing*, 14 (2): 39–41.

FCC (2000). Trends in International Telephony. Federal Communications Commission. http://www.fcc.gov.

Feinler, J. (1999). Reflecting on 30 Years of RFCs. RFC 2555: ftp://ftp.isi.edu/in-notes/rfc2555.txt.

Fischer, C. S. (1992). *America Calling: A Social History of Telephone to 1940*. Berkeley: University of California Press.

Fleck, L. (1979). *Genesis and Development of a Scientific Fact*. Chicago: The University of Chicago Press.

Foucault, M. (1970). *The Order of Things: An Archaeology of the Human Sciences*. London: Tavistock Publishers.

Freeman, C. (2000). Social inequality, technology and economic growth. In S. Wyatt, F. Henwood, N. Miller, and P. Senker (eds.), *Technology and In/equality: Questioning the Information Society* (pp. 149–71). London: Routledge.

Freeman, C., Clark, J., and Soete, L. (1982). *Unemployment and Technical Innovation: A Study of Long Waves and Economic Development*. Westport, Conn.: Greenwood Press.

Gamson, W. A. (1966). Reputation and resources in community politics. *American Journal of Sociology*, 72 (2): 121–31.

Gardner, H. (1987). *The Mind's New Science: A History of Cognitive Revolution*. New York: Basic Books.

Gibson, J. J. (1950). *The Perception of the Visual World*. Cambridge, Mass.: The Riverside Press.

—— (1979). *The Ecological Approach to Visual Perception*. Boston: Houghton Mifflin.

Giddens, A. (1990). *The Consequences of Modernity*. Cambridge: Polity Press.

Granovetter, M. (1985). Economic action and social structure: the problem of embeddedness. *American Journal of Sociology*, 91: 481–510.

Griffin, A. (1997). PDMA research on new product development practices: updating trends and benchmarking best practices. *Journal of Product Innovation Management*, 14: 429–58.

Habermas, J. (1993). *Moral Consciousness and Communicative Action*. Cambridge, Mass.: The MIT Press.

Hafner, K., and Lyon, M. (1998). *Where Wizards Stay Up Late: The Origins of the Internet*. New York: Simon & Schuster.

Halbwachs, M. (1980). *The Collective Memory*. New York: Harper & Row.

Halbwachs, M. (1992). *On Collective Memory*. Chicago: Chicago University Press.

Hamel, G. (1999). Bringing Silicon Valley inside. *Harvard Business Review*, (September–October): 71–84.

Harré, R., Clarke, D., and De Carlo, N. (1985). *Motives and Mechanisms*. London: Methuen.

Harvey, D. (1990). *The Condition of Postmodernity: An Enquiry into the Origins of Cultural Change*. Cambridge, Mass.: Blackwell.

Hauben, M., and Hauben, R. (1997). *Netizens: On the History and Impact of Usenet News and the Internet*. IEEE Computer Society Press.

Heinämaa, S., and Tuomi, I. (1989). *Ajatuksia synnyttävät koneet: tekoälyn unia ja painajaisia* (*Thought Provoking Machines: Dreams and Nightmares of Artificial Intelligence*; in Finnish). Porvoo: Werner Söderström Osakeyhtiö.

Henderson, R. M., and Clark, K. B. (1990). Architectural innovation: the reconfiguration of existing systems and the failure of established firms. *Administrative Science Quarterly*, 35: 9–30.

Himanen, P. (2001). *The Hacker Ethic and the Spirit of the Information Age*. New York: Random House.

Hofstede, G. (1991). *Cultures and Organizations: Software of the Mind*. London: McGraw-Hill.

Hughes, T. P. (1983). *Networks of Power: Electrification in Western Society 1880–1930*. Baltimore: The John Hopkins University Press.

Hugill, P. J. (1999). *Global Communications Since 1844: Geopolitics and Technology*. Baltimore: The Johns Hopkins University Press.

Janhunen, R. (1964). Näin se syntyi! *Radioamatööri*, 158–9.

Kahn, R. E., and Cerf, V. (2000). Al Gore and the Internet. http://www.cluebot.com/article.pl?sid = 00/09/29/0711253.

Kelly, P., Kranzberg, M., Rossini, F. A., Baker, N. R., Tapler, F. A., and Mitzner, M. (1978). *Technological Innovation: A Critical Review of Current Knowledge*. San Francisco: San Francisco Press.

Kendall, L. (1999). Recontextualizing 'cyberspace': methodological considerations for on-line research. In S. Jones (ed.), *Doing Internet Research: Critical Issues and Methods for Examining the Net* (pp. 57–74). Thousand Oaks, Calif.: Sage Publishers.

Kenney, M. (2000). *Understanding Silicon Valley: The Anatomy of an Entrepreneurial Region*. Stanford, Calif.: Stanford University Press.

Kirstein, P. T. (1999). Early experiences with the Arpanet and Internet in the United Kingdom. *IEEE Annals of the History of Computing*, 21 (1): 38–44.

Kleinknecht, A. (1987). *Innovation Patterns in Crisis and Prosperity: Schumpeter's Long Cycle Reconsidered*. Houndmills, Basingstoke: The Macmillan Press.

Kleinrock, L. (1961). Information Flow in Large Communication Nets, Proposal for a Ph.D. Thesis. Massachusetts Institute of Technology, Research Laboratory of Electronics. http://www.lk.cs.ucla.edu.

Kline, S. J., and Rosenberg, N. (1986). An overview of innovation. In R. Landau and N. Rosenberg (eds.), *The Positive Sum Strategy: Harnessing Technology for Economic Growth* (pp. 275–307). Washington, DC: National Academy Press.

Knorr Cetina, K. (1999). *Epistemic Cultures: How the Sciences Make Knowledge*. Cambridge, Mass.: Harvard University Press.

Kodama, F. (1995). *Emerging Patterns of Innovation: Sources of Japan's Technological Edge*. Boston: Harvard Business School Press.

Kogut, B., and Zander, U. (1997). Knowledge of the firm: combinative capabilities, and replication of technology. In L. Prusak (ed.), *Knowledge in Organizations* (pp. 17–35). Boston: Butterworth–Heinemann.

Kollock, P. (1994). The emergence of exchange structures: an experimental study of uncertainty, commitment, and trust. *American Journal of Sociology*, 100 (2): 313–45.

—— (1999). The economies of online cooperation: gifts and public goods in cyberspace. In M. A. Smith and P. Kollock (eds.), *Communities in Cyberspace* (pp. 220–39). London: Routledge.

Kozulin, A. (1990). *Vygotsky's Psychology: A Biography of Ideas*. Cambridge, Mass.: Harvard University Press.

Kramer, R. M., and Tyler, T. R. (1996). *Trust in Organizations: Frontiers of Theory and Research*. Thousand Oaks, Calif.: Sage Publications.

Krum, H. L. (1925). A Brief History of the Morkrum Company. http://www.funet.fi/pub/telecom/telecom-archives/history/teletype.gz.

Kuhn, T. S. (1970). *The Structure of Scientific Revolutions*. Chicago: The University of Chicago Press.

—— (1979). Foreword. In *Ludwik Fleck: Genesis and Development of a Scientific Fact* (pp. vii–xi). Chicago: Chicago University Press.

Kusnetzky, D., and Gillen, A. (2001). Linux: A Journey into the Enterprise: An IDC White Paper Sponsored by Red Hat. IDC. http://www.redhat.com/whitepapers/.

Kuwabara, K. (2000). Linux: a bazaar at the edge of chaos. *First Monday*, 5 (3). http://first-monday.org/issues/issue5_3/kuwabara/.

Lakoff, G. (1987). *Women, Fire, and Dangerous Things: What Categories Reveal about the Mind*. Chicago: University of Chicago Press.

Lang, G. E., and Lang, K. (1988). Recognition and renown: the survival of artistic reputation. *American Journal of Sociology*, 94 (1): 79–109.

Latour, B. (1999). *Pandora's Hope: Essays on the Reality of Science Studies*. Cambridge, Mass.: Harvard University Press.

—— and Woolgar, S. (1986). *Laboratory Life: The Construction of Scientific Facts*. Princeton: Princeton University Press.

Lave, J., and Wenger, E. (1991). *Situated Learning: Legitimate Peripheral Participation*. Cambridge: Cambridge University Press.

Law, J. (1992). Note on the Theory of the Actor-Network: Ordering, Strategy, and Heterogeneity. *Systems Practice*, 5 (4): 379–93.

—— and Hassard, J. (1999). *Actor Network Theory and After*. Oxford: Blackwell.

Lee, J. A. N. (1992). Claims to the term 'time sharing'. *IEEE Annals of the History of Computing*, 14 (1): 16–17.

Lee, J. A. N., Fano, R. M., Scherr, A. L., Corbató, F., and Vyssotsky, V. A. (1992). Project MAC (time-sharing computing project). *IEEE Annals of the History of Computing*, 14 (2): 9–13.

Leiner, B. M., Cerf, V. G., Clark, D. D., Kahn, R. E., Kleinrock, L., Lynch, D. C., Postel, J., Roberts, L. G., and Wolff, S. (2000). A Brief History of the Internet, version 3.31, 4 Aug 2000. http://www.isoc.org/internet-history/brief.html.

Leont'ev, A. N. (1978). *Activity, Consciousness, and Personality*. Englewood Cliffs, NJ: Prentice Hall.

Lessig, L. (1999). *Code: and Other Laws of Cyberspace*. New York: Basic Books.

Levinas, E. (1969). *Totality and Infinity*. Pittsburgh: Duquesne University Press.

Levy, S. (1984). *Hackers: Heroes of the Computer Revolution*. Garden City, NY: Anchor Press/Doubleday.

Lewicki, R. J., and Bunker, B. B. (1996). Developing and maintaining trust in work relationships. In R. M. Kramer and T. R. Tyler (eds.), *Trust in Organizations: Frontiers of Theory and Research* (pp. 114–39). Thousand Oaks, Calif.: Sage Publications.

Licklider, J. C. R. (1960). Man-Computer Symbiosis. *IRE Transactions on Human Factors in Electronics*, HFE-1, pp. 4–11. Reprinted: In Memoriam: J. C. R. Licklider (1915–1990), Digital Systems Research Center, 7 August 1990. ftp://ftp.digital.com/pub/DEC/SRC/research-reports/.

—— and Taylor, R. W. (1968). The computer as a communication device. *Science and Technology* (April). Reprinted: In Memoriam: J. C. R. Licklider (1915–1990), Digital Systems Research Center, 7 August 1990. ftp://ftp.digital.com/pub/DEC/SRC/research-reports/.

Luhmann, N. (1990). *Essays on Self-Reference*. New York: Columbia University Press.

—— (1995). *Social Systems*. Stanford, Calif.: Stanford University Press.

Lukka, T. J. (2001). A Gentle Introduction to Ted Nelson's ZigZag Structure. http://gzigzag. sourceforge.net/gi/gi-ns4.html.

Luria, A. R., and Vygotsky, L. (1992). *Ape, Primitive Man, and Child: Essays in the History of Behavior*. Hemel Hempstead: Harvester Wheatsheaf.

Lynn, G. S., Morone, J. G., and Paulson, A. S. (1997). Marketing and discontinuous innovation: the probe and learn process. In M. L. Tushman and P. Anderson (eds.), *Managing Strategic Innovation and Change: A Collection of Readings* (pp. 353–76). Oxford: Oxford University Press.

Lynn, L. H., Aram, J. D., and Reddy, N. M. (1997). Technology communities and innovation communities. *Journal of Engineering and Technology Management*, 14: 129–45.

McCorduck, P. (1979). *Machines Who Think: A Personal Inquiry into the History and Prospects of Artificial Intelligence*. San Francisco: W. H. Freeman and Company.

McCracken, G. (1988). *Culture and Consumption: New Approaches to the Symbolic Character of Consumer Goods and Activities*. Bloomington: Indiana University Press.

McCullagh, D. (1999). No Credit Where It's Due. *Wired*, 11 Mar. 1999; http://www.wired.com/ news/politics/0,1283,18390,00.html.

—— (2000). The Mother of Gore's Invention. *Wired*, 17 Oct. 2000; http://www.wired.com/ news/politics/0,1283,39301,00.html.

Machlup, F., and Penrose, E. (1950). The patent controversy in the nineteenth century. *Journal of Economic History*, 10 (1): 1–29.

MacIntyre, A. (1981). *After Virtue: A Study in Moral Theory*. London: Duckworth.

McKee, D. (1992). An organizational learning approach to product innovation. *Journal of Product Innovation Management* 9: 232–45.

McKusick, M. K. (1999). Twenty years of Berkeley Unix: from AT&T-owned to freely redistributable. In C. DiBona, S. Ockman, and M. Stone (eds.), *Open Sources: Voices of the Open Source Revolution* (pp. 31–46). Sebastopol, Calif.: O'Reilly & Associates.

Mahajan, V., and Wind, J. (1992). New product models: practice, shortcomings and desired improvements. *Journal of Product Innovation Management*, 9: 128–39.

Mandel, E. (1995). *Long Waves of Capitalist Development: A Marxist Interpretation* (revised edn.). London: Verso.

Maturana, H. R., and Varela, F. J. (1980). *Autopoiesis and Cognition: The Realization of the Living*. London: Reidl.

—— and Varela, F. J. (1988). *The Tree of Knowledge: The Biological Roots of Human Understanding*. Boston: New Science Library.

Merton, R. K. (1973). *The Sociology of Science*. Chicago: University of Chicago Press.

—— (1988). The Matthew Effect in science, II: Cumulative advantage and the symbolism of intellectual property. *Isis*, 79 (299): 606–23.

—— (1995). The Thomas Theorem and the Matthew Effect. *Social Forces*, 74 (2): 379–401.

Meyerson, D., Weick, K. E., and Kramer, R. M. (1996). Swift trust and temporary groups. In R. M. Kramer and T. R. Tyler (eds.), *Trust in Organizations: Frontiers of Theory and Research* (pp. 166–95). Thousand Oaks, Calif.: Sage Publications.

Miles, R. E., Snow, C. C., Mathews, J. A., Miles, G., and Coleman, H. J., Jr. (1997). Organizing in the knowledge age: Anticipating the cellular form. *Academy of Management Executive*, 11 (4): 9–24.

Mintzberg, H. (1979). *The Structuring of Organizations*. Englewood Cliffs, NJ: Prentice-Hall.

Mitchell, W. J. (1995). *City of Bits: Space, Place, and the Infobahn*. Cambridge, Mass.: The MIT Press.

Mokyr, J. (1990). *The Lever of Riches: Technological Creativity and Economic Progress*. Oxford: Oxford University Press.

Morson, G. S., and Emerson, C. (1990). *Mikhail Bakhtin: Creation of Prosaics*. Stanford, Calif.: Stanford University Press.

Mullins, J. W., and Sutherland, D. J. (1998). New product development in rapidly changing markets: an exploratory study. *Journal of Product Innovation Management*, 15: 224–36.

Mundie, G. (2001). The commercial software model. http://www.microsoft.com/presspass/exec/craig/05–03sharedsource.asp.

Nahapiet, J., and Ghoshal, S. (1998). Social capital, intellectual capital, and the organizational advantage. *Academy of Management Review*, 23 (2): 242–66.

Nardi, B. A., Whittaker, S., and Schwartz, H. (2000). It's Not What You Know, It's Who You Know: Work in the Information Age. *First Monday*, 5 (5). http://firstmonday.org/issues/issue5_5/nardi/.

Naughton, J. (2000). *A Brief History of the Future: From Radio Days to Internet Years in a Lifetime*. Woodstock: The Overlook Press.

Nelson, R. A. (1963). History of Typewriter Development. Teletype Corporation, 1963; available at http://www.funet.fi/pub/doc/telecom/telecom-archives/history/teletype.gz.

Nelson, R. R. (1994). Intellectual property protection for cumulative systems technology. *Columbia Law Review*, 94: 2674–7.

—— and Winter, S. G. (1977). In search of useful theory of innovation. *Research Policy*, 6: 36–76.

Nelson, T. H. (1995*a*). The heart of connection: hypermedia unified by transclusion. *Communications of the ACM*, 38 (8): 31–3.

—— (1995*b*). Transcopyright: pre-premission for virtual republishing. http//www.sfc.keio.ac.jp/~ted/transcopyright/transcopy.html.

—— (1998). What's on my mind: an invited talk at the first Wearable Computer Conference, Fairfax Va., 12–13 May 1998. http://www.sfc.keio.ac.jp/~ted/zigzag/ xybrap.html.

Nishida, K. (1920). The Standpoint of Religion. http://www.aloha.net/~albloom/sdn/nishida1.htm.

—— (1958). *Intelligibility and the Philosophy of Nothingness: Three Philosophical Essays*. Honolulu: East-West Center.

—— (1987*a*). *Intuition and Reflection in Self-Consciousness*. New York: State University of New York Press.

—— (1987*b*). *Last Writings: Nothingness and the Religious Worldview*. Honolulu: University of Hawaii Press.

—— (1990). *An Inquiry into the Good* (orig. pub. 1911). New Haven: Yale University Press.

Nishitani, K. (1991). *Nishida Kitaro*. Berkeley: University of California Press.

Nonaka, I. (1988). Speeding organizational information creation: toward middle-up-down management. *Sloan Management Review* (Spring): 57–73.

—— (1994). A dynamic theory of organizational knowledge creation. *Organization Science*, 5: 14–37.

—— and Konno, N. (1998). The concept of 'ba': building a foundation for knowledge creation. *California Management Review*, 40 (3): 40–54.

—— and Takeuchi, H. (1995). *The Knowledge-Creating Company: How Japanese Companies Create the Dynamics of Innovation*. Oxford: Oxford University Press.

—— Toyama, R., and Konno, N. (2000). SECI, ba, and leadership: a unified model of dynamic knowledge creation. *Long Range Planning*, 33: 5–34.

—— and Nagata, A. (2000). A firm as a knowledge-creating entity: a new perspective on the theory of the firm. *Industrial and Corporate Change*, 9 (1): 1–20.

Norberg, A. L. (1996). Changing computing: the computing community and DARPA. *IEEE Annals of the History of Computing*, 18 (2): 40–53.

NSF (2000). Science and Engineering Indicators: Chapter I, Science and Technology in Transition: the 1940s and 1990s. National Science Foundation. http://www.nsf.gov/.

Olmstedt, D. D. (1998). History and Principles of Neural Networks. http://www.neurocomputing.org/history.htm.

O'Neill, J. E. (1995). The role of ARPA in the Development of the ARPANET, 1961–1972. *IEEE Annals of the History of Computing*, 17 (4): 76–81.

Pacey, A. (1999). *Meaning in Technology*. Cambridge, Mass.: The MIT Press.

Padmore, T., Schuetze, H., and Gibson, H. (1998). Modeling systems of innovation: an enterprise-centered view. *Research Policy*, 26: 605–24.

Pam, A. (2000). Where World Wide Web Went Wrong. http://xanadu.com.au/xanadu/6w-paper.html, retrieved 11 October 2000.

Peeling, N., and Satchell, J. (2001). Analysis of the Impact of Open Source Software. QinetiQ Ltd. QINETIQ/KI/SEB/CR010223. http://www.govtalk.gov.uk/interoperability/egif_document.asp?docnum = 430.

Perens, B. (1999). The Open Source Definition. In C. DiBona, S. Ockman, and M. Stone (eds.), *Open Sources: Voices from the Open Source Revolution* (pp. 171–88). Sebastopol, Calif.: O'Reilly & Associates, Inc.

Perez, C. (1985). Microelectronics, long waves and world structural change: new perspectives for developing countries. *World Development*, 13 (3): 441–63.

Petroski, H. (1994). *The Evolution of Useful Things: How Everyday Artifacts—From Forks and Pins to Paper Clips and Zippers—Came to Be as They Are*. New York: Vintage Books.

—— (1996). *Invention by Design: How Engineers Get from Thought to Thing*. Cambridge, Mass.: Harvard University Press.

Polanyi, M. (1967). *The Tacit Dimension*. New York: Anchor.

—— (1998). *Personal Knowledge: Towards a Post-Critical Philosophy*. London: Routledge.

—— and Prosch, H. (1975). *Meaning*. Chicago: The University of Chicago Press.

Powell, W. W., Koput, K. W., and Smith-Doerr, L. (1996). Interorganizational collaboration and the locus of innovation: networks of learning in biotechnology. *Administrative Science Quarterly*, 41 (1): 116–45.

Quarterman, J. S. (1999). Revisionist internet history. *Matrix News*, 9 (4). http://www.mids.org/mn/904/.

RAND (2000). Fifty Years of Service to the Nation. http://www.rand.org/50TH/.

Rawls, J. (1999). *The Law of Peoples: with 'The Idea of Public Reason Revisited'*. Cambridge, Mass.: Harvard University Press.

Raymond, E. R. (1998a). Homesteading the noosphere. *First Monday*, 3 (10). http://firstmonday.org/issues/issue3_10/raymond/, and http://www.tuxedo.org/-esr/writings/.

—— (1998b). The Cathedral and the Bazaar. *First Monday*, 3 (3). http://firstmonday.org/issues/issue3_3/raymond/, and http://www.tuxedo.org/-esr/writings/.

—— (1999). *The Cathedral and the Bazaar: Musings on Linux and Open Source by an Accidental Revolutionary*. Sebastopol, Calif.: O'Reilly & Associates, Inc.

Reid, E. (1999). Hierarchy and power: social control in cyberspace. In M. A. Smith and P. Kollock (eds.), *Communities in Cyberspace* (pp. 107–33). London: Routledge.

Reid, R. H. (1997). *Architects of the Web: 1,000 Days that Built the Future of Business*. New York: John Wiley & Sons.

Rheingold, H. (1993). *The Virtual Community: Homesteading on the Electronic Frontier*. Reading, Mass.: Addison-Wesley.

Roberts, L. G. (1999a). Internet Chronology. http://www.ziplink.net/~lroberts/Internet Chronology.html.

—— (1999b). Packet Switching—Early Development. Speech delivered at SIGCOMM 99. http://www.ziplink.net/-lroberts/internet_history:information.html.

Rogers, E. M. (1995). *Diffusion of Innovations* (4th edn.). New York: The Free Press.

Rogoff, B. (1990). *Apprenticeship in thinking: cognitive development in social contexts.* New York: Oxford University Press.

Rosen, R. (2000). *Essays on Life Itself.* New York: Columbia University Press.

Rosenbloom, R. S., and Christensen, C. M. (1994). Technological discontinuities, organizational capabilities, and strategic commitments. *Industrial and Corporate Change,* 3: 655–85.

Samuelson, P., Davis, R., Kapor, M., and Reichman, J. H. (1994). A manifesto concerning the legal protection of computer programs. *Columbia Law Review,* 94: 2308–431.

—— and Glushko, R. J. (1993). Intellectual property rights for digital library and hypertext publishing systems. *Harvard Journal of Law and Technology,* 6: 237–61.

Sawhney, M., and Prandelli, E. (2000). Communities of creation: managing distributed innovation in turbulent markets. *California Management Review,* 42 (2): 24–54.

Saxenian, A. (1994). *Regional Advantage: Culture and Competition in Silicon Valley and Route 128.* Cambridge, Mass.: Harvard University Press.

Schön, D. A. (1963). *Invention and the Evolution of Ideas.* London: Social Science Paperbacks.

—— (1983). *The Reflective Practitioner.* New York: Basic Books.

Schumpeter, J. A. (1975). *Capitalism, Socialism and Democracy.* New York: Harper and Row.

Shapin, S. (1995). Here and everywhere: sociology of scientific knowledge. *Annual Review of Sociology,* 21: 289–321.

Shimizu, H., and Yamaguchi, Y. (1987). Synergetic computer and holonics: information dynamics of a semantic computer. *Physica Scripta,* 36 (6): 970–85.

Shirky, C. (2001). Listening to Napster. In A. Oram (ed.), *Peer-To-Peer: Harnessing the Benefits of a Disruptive Technology* (pp. 21–37). Sebastopol, Calif.: O'Reilly & Associates, Inc.

Simmel, G. (1990). *The Philosophy of Money* (2nd enlarged edn.; 1st edn. 1900). London: Routledge.

Smith, J. B., and Weiss, S. F. (1988). Hypertext. *Communications of the ACM,* 31 (7): 816–19.

Soper, B., Milford, G., and Rosenthal, G. (1995). Belief when evidence does not support theory. *Psychology and Marketing,* 12 (5): 415–22.

Spender, J.-C. (1996). Making knowledge as the basis of a dynamic theory of the firm. *Strategic Management Journal,* 17 (Special Issue): 45–62.

Spender, J.-C. (1998). The dynamics of individual and organizational knowledge. In C. Eden and J.-C. Spender (eds.), *Managerial and Organizational Cognition: Theory, Methods and Research.* London: Sage Publications.

Stallman, R. (1999). The GNU operating system and the free software movement. In C. DiBona, S. Ockman, and M. Stone (eds.), *Open Sources: Voices from the Open Source Revolution* (pp. 53–70). Sebastopol, Calif.: O'Reilly & Associates, Inc.

Standage, T. (1998). *The Victorian Internet: The Remarkable Story of the Telegraph and the Nineteenth Century's On-line Pioneers.* New York: Walker Publishing Company.

Stankiewicz, R. (2000). The concept of 'design space'. In J. Ziman (ed.), *Technological Innovation as an Evolutionary Process* (pp. 234–47). Cambridge: Cambridge University Press.

Stigler, S. M. (1999). *Statistics on the Table: The History of Statistical Concepts and Methods.* Cambridge, Mass.: Harvard University Press.

Takeuchi, Y. (1963). The philosophy of Nishida. *Japanese Religions,* 3 (4): 11–17.

Tanenbaum, A. S., and Woodhull, A. (1997). *Operating Systems: Design and Implementation* (2nd edn.). Upper Saddle River, NJ: Prentice Hall.

Taylor, C. (1989). *Sources of the Self: The Making of the Modern Identity*. Cambridge: Cambridge University Press.

Taylor, R. W. (1990). Preface. In Memoriam: J. C. R. Licklider, 1915–1990. Digital Systems Research Center. ftp://ftp.digital.com/pub/DEC/SRC/research-reports/.

Teece, D. J., Pisano, G., and Shuen, A. (1997). Dynamic capabilities and strategic management. *Strategic Management Journal*, 18 (7): 509–33.

The Times (2000). Donald Davies (obituary). *The Times*, http://www.the-times.co.uk/news/pages/tim/2000/05/31/timobiobio2004.html.

Thomke, S., Von Hippel, E., and Franke, R. (1998). Modes of experimentation: an innovation process—and competitive—variable. *Research Policy*, 27: 315–32.

Thompson, C. J., and Haytko, D. L. (1997). Speaking of fashion: consumers' uses of fashion discourses and the appropriation of countervailing cultural meanings. *Journal of Consumer Research*, 24 (1): 15–42.

Torvalds, L. (1999). The Linux edge. In C. DiBona, S. Ockman, and M. Stone (eds.), *Open Sources: Voices from the Open Source Revolution* (pp. 101–11). Sebastopol, Calif.: O'Reilly & Associates, Inc.

—— (2000). Linux-kernel mailing list summary, Kernel Traffic #60, 27 Mar, 2000. http://kt.linuxcare.com/.

—— (2001). What makes hacker stick? a.k.a. Linus's Law. In P. Himanen (ed.), *The Hacker Ethic and the Spirit of the Information Age* (pp. xiii–xvii). London: Secker & Warburg.

Tuomi, I. (1982). Elämän synnystä (*On the origin of life*). *Arkhimedes*, 34: 106–16.

—— (1987). *Ei ainoastaan hakkerin käsikirja (Not Only a Hacker's Handbook)*. Porvoo: WSOY.

—— (1996). The communicative view on organizational memory: power and ambiguity in knowledge creation systems. In J. F. Nunamaker, Jr., and R. H. Sprague, Jr. (eds.), *Proceedings of the Twenty-Ninth Hawaii International Conference on System Sciences* (pp. 147–55). Los Alamitos, Calif.: IEEE Computer Society Press.

—— (1998). Vygotsky in a TeamRoom: an exploratory study on collective concept formation in electronic environments. In J. F. Nunamaker, Jr. (ed.), *Proceedings of the 31st Annual Hawaii International Conference on System Sciences* (pp. 68–75). Los Alamitos, Calif.: IEEE Computer Society Press.

—— (1999). *Corporate Knowledge: Theory and Practice of Intelligent Organizations*. Helsinki: Metaxis.

—— (2000). Data is more than knowledge: implications of the reversed knowledge hierarchy to knowledge management and organizational memory. *Journal of Management Information Systems*, 6 (3): 103–17.

—— (2001). Internet, innovation, and open source: actors in the network. *First Monday*, 6 (1). http://firstmonday.org/issues/issue6_1/tuomi/index.html.

Tushman, M. L., and Anderson, P. (1986). Technological Discontinuities and Organizational Environments. *Administrative Science Quarterly*, 31 (September): 439–65.

US Census Bureau (1999). Historical Census of Housing Tables—Telephones. http://www.census.gov/hhes/www/housing/census/historic/phone.html.

Urban, G., and Von Hippel, E. (1988). Lead user analyses for the development of new industrial products. *Management Science*, 34 (5): 569–82.

Usher, A. P. (1954). *History of Mechanical Inventions*. Boston: Harvard University Press.

Utterback, J. M. (1994). *Mastering the Dynamics of Innovation: How Companies Can Seize Opportunities in the Face of Technological Change*. Boston: Harvard Business School Press.

—— and Abernathy, W. J. (1976). A dynamic model of process and product innovation. *Omega*, 3 (6): 639–56.

Van Bragt, J. (1982). Translator's introduction. In *Religion and Nothingness by Keiji Nishitani* (pp. xxiii–xlv). Berkeley: University of California Press.

Van de Ven, A. H. (1993). A community perspective on the emergence of innovations. *Journal of Engineering and Technology Management*, 10: 23–51.

van der Veer, R., and Valsiner, J. (1994). *Understanding Vygotsky: A Quest for Synthesis*. Cambridge, Mass.: Blackwell.

Varian, H. R. (2000). The law of recombinant growth. *The Industry Standard*, issue 6 Mar. 2000; available at http:/www.thestandard.com/.

Verganti, R. (1999). Planned flexibility: linking anticipation and reaction in product development projects. *Journal of Product Innovation Management*, 16: 363–76.

Victor, B., and Boynton, A. C. (1998). *Invented Here: Maximizing Your Organization's Internal Growth and Profitability*. Boston, Mass.: Harvard Business School Press.

Von Hippel, E. (1976). The dominant role of users in the scientific instrument innovation process. *Research Policy*, 5 (3): 212–39.

—— (1988). *The Sources of Innovation*. New York: Oxford University Press.

von Krogh, G., Ichijo, K., and Nonaka, I. (2000). *Enabling Knowledge Creation: How to Unlock the Mystery of Tacit Knowledge and Release the Power of Innovation*. Oxford: Oxford University Press.

Vygotsky, L. (1978). *Mind in Society: The Development of Higher Psychological Processes*. Cambridge, Mass.: Harvard University Press.

Wahba, A., and Bridgewell, L. (1976). Maslow reconsidered: a review of research on the need hierarchy theory. *Organizational Behavior and Human Performance*, 15: 212–40.

Watson, P. (1976). *Building the Medieval Cathedral*. Minneapolis: Lerner Publications Company.

Wayner, P. (2000). *Free for All: How Linux and the Free Software Movement Undercut the High-Tech Titans*. New York: HarperBusiness.

Weizenbaum, J. (1984). *Computer Power and Human Reason: From Judgement to Calculation*. Harmondsworth: Penguin Books.

Wenger, E. (1998). *Communities of Practice: Learning, Meaning, and Identity*. Cambridge: Cambridge University Press.

Wertsch, J. V. (1981). *The Concept of Activity in Soviet Psychology*. Armonk, NY: Sharpe.

—— (1985). *Vygotsky and the Social Formation of Mind*. Cambridge, Mass.: Harvard University Press.

—— (1991). *Voices of the Mind: A Sociocultural Approach to Mediated Action*. Cambridge, Mass.: Harvard University Press.

—— (1998). *Mind as Action*. Oxford: Oxford University Press.

—— del Río, P., and Alvarez, A. (1995). Sociocultural studies: history, action, and mediation. In J.V. Wertsch, P. del Río, and A. Alvarez (eds.), *Sociocultural Studies of Mind* (pp. 1–34). Cambridge: Cambridge University Press.

Wiener, N. (1975). *Cybernetics: or Control and Communication in the Animal and the Machine*. Cambridge, Mass.: The MIT Press.

Wilson, R. (1985). Reputations in games and markets. In A. E. Roth (ed.), *Game-Theoretic Models of Bargaining* (pp. 27–62). Cambridge: Cambridge University Press.

Ziman, J. (2000). Selectionism and complexity. In J. Ziman (ed.), *Technological Innovation as an Evolutionary Process* (pp. 41–51). Cambridge: Cambridge University Press.

Zuckerman, H. (1988). Intellectual property and diverse rights of ownership in science. *Science, Technology, and Human Values*, 13 (1&2): 7–16.

Name Index

Subject Index